SpringerBriefs in Earth Sciences

More information about this series at http://www.springer.com/series/8897

Robert J. Brewer

Hydrocarbon Potential
in Southeastern United States

A Review

 Springer

Robert J. Brewer
Katy, TX, USA

ISSN 2191-5369   ISSN 2191-5377   (electronic)
SpringerBriefs in Earth Sciences
ISBN 978-3-030-00216-9   ISBN 978-3-030-00218-3   (eBook)
https://doi.org/10.1007/978-3-030-00218-3

Library of Congress Control Number: 2018955281

This Springer imprint is published by the registered company Springer Nature Switzerland AG
The registered company address is: Gewerbestrasse 11, 6330 Cham, Switzerland

*I dedicate this work to the memory
of my parents – Ralph G. Brewer
and Elizabeth A. Brewer. Their steadfast
love, encouragement, and support always
sustained me.*

Soli Deo Gloria

Disclaimer Notice

The author, Robert J. Brewer, Managing Director of R.J. Brewer and Associates, LLC, has a written agreement executed on December 4, 2014, to publish this work with Springer International Publishing AG., New York. The author is in the process of preparing the final manuscript and producing figure illustrations that may appear in the final published version.

The views and opinions expressed in this work are those of the author only. This is an historical writing with indications to possible future outcomes. References made to people, places, entities, and other things are meant only to reinforce the author's message and assist the reader's understanding in an effort to present relevant and supporting information on the subject matter of the work. References and conclusions that may be drawn by any individual or individuals that may result from information presented in the work that may be construed to have a bearing on any person, place, or thing are entirely coincidental.

Acknowledgments

I thank the following family members, friends, and colleagues for their support and encouragement during the production of this work: my wife – Brenda Brewer for her patience, encouragement, and helpful suggestions with the writing and illustrations; my son – Brett Brewer for his support and encouragement; my sisters Martha Rini and Joan Roman for their encouragement and support; my editor – Ron Doering of Springer Science for his foresight, guidance, patience, and encouragement throughout the process from beginning to end; Tom Harrold for his friendship, suggestions, and encouragement; Jim Kane for his friendship, valuable insights about the geology of the Gulf Coast, expertise, suggestions, and encouragement; Walt Huff for his encouragement, expertise, and valuable insights about the state of Florida; and Jim Kennedy for his untiring help and attention to detail in providing well log and historical information from the state of Georgia.

Contents

List of Figures

About the Author

Robert James Brewer is the Managing Director and Owner of R.J. Brewer and Associates, LLC which is an oil and gas exploration services sales, business development, technical writing and electronic commerce corporation.

He has been actively involved in the oil and gas exploration business since 1979 and is an experienced technical sales and operations management professional. He is an effective goal-oriented team builder, leader and player. He has done nearly everything one can reasonably do with seismic data including data acquisition, processing, interpretation and mapping as well as seismic data sales and trades. He has been involved in numerous successful seismic data acquisition and processing projects - both surface and borehole throughout the lower 48 states, Canada and Mexico. He has large independent and major oil company work experience starting out in computer operations with the Standard Oil Company of Ohio (Sohio), then moving on into seismic data acquisition and processing at the Geophysical Service Inc. (GSI) division of Texas Instruments, Inc. He then advanced into exploration geophysics with Texaco Inc. He and his teams mapped large portions of the Permian Basin of Texas and New Mexico recommending the drilling of several deep wildcat and development well locations.

He is considered to be an established subject matter resource on surface and borehole seismic (VSP) data acquisition and processing project planning, execution, sales and business development. He was instrumental with focused sales efforts starting and building successful surface and bore hole seismic and micro seismic data acquisition and processing operations at large integrated public oil and gas service companies including Halliburton and later at Baker Hughes Inc. He is a published technical writer as well as a successful entrepreneur in electronic commerce (e-commerce).

He earned a BA degree in geology with a minor in history from Bowling Green State University (BGSU). He is married and resides in the Houston, Texas area.

The author's other publications include:

VSP is a Check Shot Step Up, 2000, AAPG Explorer
The Check Shot Velocity Survey, 2000, AAPG Explorer
The Look ahead VSP, 2002, AAPG Search and Discovery
Comprehensive Use of VSP Technology at Elk Hills Field, Kern County, CA., 2003, AAPG Search and Discovery, co-author
Micro Seismic Fracture Monitoring, 2006, SPWLA.
Applications of Vertical Seismic Profile, 2007, SPWLA.
Micro Seismic Monitoring of a Permian Basin San Andres Dolomite Horizontal Well, co-author, SPE, 2007
New Technologies and Applications of the Vertical Seismic Profile as a Litho-Structural Tool and "Look Ahead" Mode, co-author, SPE 2007.
Look Ahead Applications of Vertical Seismic Profiles as a Litho Structural Tool combined with Dipole Sonic Logging: Case Histories - Burgos Basin- Northern Mexico, co-author., SPE 2007.

Chapter 1
Introduction

The southeastern United States as defined by the author includes the states of Virginia, North Carolina, South Carolina, Georgia and extends south through the Florida peninsula. The relevance of the states bordering the area will also be discussed. The region is currently lacking significant oil and gas production except for far western Virginia, the panhandle of Florida and far southern Florida near the Everglades. Documented oil and gas shows and evidence of hydrocarbon existence have been reported in notable wells drilled in all of the areas. Small oil seeps reaching the surface have been observed in central Georgia dating back to the first settlements of the eighteenth century and were probably observed earlier before the reported and local newspaper documented discovery in the early part of the twentieth century in 1919.

Oil and gas seeps are natural springs where liquid and gaseous hydrocarbons (hydrogen-carbon compounds) leak out of the ground. Whereas freshwater springs are fed by underground pools of water, oil and gas seeps are fed by natural underground accumulations of oil and natural gas.

The portion of the southeastern United States that is believed to be prospective with undiscovered accumulations of oil and gas hydrocarbons encompasses over 1110 thousand square kilometers and is considered part of the East Coastal or Atlantic Coastal Basin. The area appears to extend as far north as Massachusetts and also includes a significant amount of area offshore. The states of Florida and Georgia are where the interpreted boundaries of the East Coastal (Atlantic Coastal) Basin and the hydrocarbon-rich Gulf of Mexico Basin converge. Exactly where the extensive boundary occurs between the two basins is unknown, has to be generalized and is subject to geological opinion and data interpretation (Fig. 1.1).

The author's interest in the hydrocarbon potential of the southeastern United States was first stimulated and started in earnest in 2004 when he purchased and studied a copy of a COCORP (the Consortium for Continental Reflection Profiling) Seismic Reflection Traverse Across the Southern Appalachians, (SRTSA). The 2D

© The Author(s), under exclusive licence to Springer Nature Switzerland AG 2019
R. J. Brewer, *Hydrocarbon Potential in Southeastern United States*,
SpringerBriefs in Earth Sciences, https://doi.org/10.1007/978-3-030-00218-3_1

Major North American Oil & Gas Basins

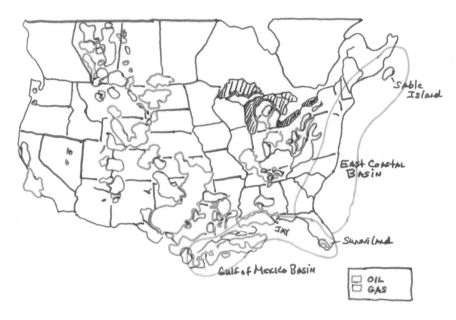

Fig. 1.1 Major North American oil and gas basins and Gulf of Mexico and East coastal basins
The East coastal basin includes areas onshore and offshore and on trend with established hydrocarbon production in south Florida on up to the far northeast at Sable Island

surface seismic data atlas and interpretation covers parts of northeastern Georgia, western North Carolina and eastern Tennessee.

Seismic data as used in oil and gas exploration involves recording the echoes or reflections bounced off rock layers far underground initiated from an artificially produced energy source activation at the surface. Analysis of the reflections after data processing helps geoscientists map the rocks in the subsurface and aids in the location of rock layers that may contain oil and/or natural gas. The reflection seismic data method was demonstrated successfully using simple energy source and recording equipment by today's standards, by a team led by J.C. Karcher (1894–1978) on a project targeting the subsurface rock formations of the Sylvan shale and Viola sandstone near Belle Isle that is located on the northwest side of Oklahoma City, in Oklahoma County, Oklahoma in 1921.

Karcher is credited with inventing and later commercializing the reflection seismograph method. He applied for patents in 1919 and created the means by which a large portion of the world's oil and gas reserves have been discovered. In 1930 he and Eugene McDermott (1899–1973) founded Geophysical Service Incorporated (GSI), a pioneering provider of seismic exploration services to the oil and gas exploration industry. In the first of series of early day divestitures; GSI later spun off the seismic exploration services business to become what later would be known as (TI) or Texas Instruments Incorporated.

I obtained the COCORP seismic data atlas during attendance at an American Association of Petroleum Geologists (AAPG) Convention in Tulsa, Oklahoma. The atlas is part of the classic COCORP project of the 1980's conducted in readily accessible areas and primarily established structural basins, some of which produce hydrocarbons such as the Texas gulf coast of the United States. The more easily accessible areas for the COCORP seismic survey project included highway and possibly railroad right of ways. One of the projects main objectives was to study plate tectonics and image crystalline basement and sub-basement rocks. The deep (over 15,000 ft.), surface seismic reflection profiling project used long record lengths with over a 12 s listen (recording) time. The project used the 2D reflection seismic profiling method and utilized truck mounted vibrators as the required energy source. Most of the data was recorded and processed by the Seismograph Service Corporation (SSC) based in Tulsa, Oklahoma. SSC was one of the author's former employers and was one of the world's oldest and largest seismograph contractors in existence at the time. SSC eventually went out of business due in part to the nearly incessant turmoil experienced by the U.S. oil and gas exploration business during the mid to late 1980's. As of this writing, the oil and gas exploration and development business in the United States as well as Canada and to a lesser extent – globally, is experiencing the worst downturn in the last 37 years as evidenced by the precipitous drop in active drilling rigs published daily by Baker Hughes. It was also especially interesting to also note that the international drilling rig rate has not decreased nearly as much as the U.S. and Canadian rig rate drop.

It was even more astonishing to note the negative difference in drilling activity from 2014 to 2015. The author's frame of reference started in 1979 when he began his oil and gas exploration career as an operations engineer and seismologist working for the long established large international seismograph contractor that was known as Geophysical Service Inc. (GSI), GSI was the parent company and later a subsidiary of the famous electronics pioneer and manufacturer known as (TI) Texas Instruments, Inc.

At the time of this writing, the U.S. oil and gas exploration industry again unfortunately finds itself in yet another very deep and serious recession period that started in late 2013 and caused in large part by an over-supply of crude oil on global markets. Due in part to the late 1980's industry downturn, SSC and GSI assets were summarily and eventually absorbed via a long process of acquisition and merger transactions into the large Schlumberger entity which is currently maintaining its place as the largest and most integrated world-wide oil and gas exploration services and industry leader in the world.

The severe downturn of the United States oil and gas exploration business that probably started in 2013 or perhaps earlier, resulted in the loss of at least a quarter of a million high-paying oil and gas industry jobs in the United States alone. Some estimates put the figure much higher and closer to 500,000. The social and economic losses in the United States has been substantial and resulted in billions of dollars being removed from the U.S. economy. The actual lost job total is probably much higher when one considers global job loss and supporting industry job losses. For the sake of simplicity and presuming that each job represented net annual earnings of $50,000 per year, the amount of money taken out of the U.S. economy involving 250,000 lost jobs is $12.5 billion. A reality in addition to the lost jobs that seems to get lost with a large part of the American public and especially some politicians is that the loss of the oil and gas jobs has also produced a huge ripple effect throughout the economy nation-wide producing large losses to other American businesses and jobs that relied on income from the decimated oil and gas business. The oil and gas industry depression that started in earnest in 2013 appears to be worse than the severe depression of the industry that started in the early 1980's. What the two problems have in common is the price drops that occurred with the price of crude oil. In the early 1980's episode, the price of oil went down from $100 per barrel in 1981 to below $20.00 per barrel in 1999. In the episode that seemed to get well underway in 2013, the price of oil quickly dropped from over $100 per barrel in 2011 to below as $40.00 per barrel in 2014. The price drop events were rapid and ended up decimating the oil and gas industries by costing many thousands of industry employees their livelihoods. The job losses also created a huge ripple effect worldwide as support industries saw their revenues rapidly decrease resulting in even more job losses. The problem is not new and also has occurred in other commodity based businesses like lumber and steel, but the issues with the fluctuating prices of oil and natural gas are more widespread across the economy as they more significantly effect a society's power generation, manufacturing and transportation costs.

The severe domestic oil and gas industry down turn or depression that started in 2013 is believed to have been caused by at least three significant occurrences. The occurrences include an oversupply of U.S. produced oil and gas resulting from the oil and gas exploration success of American oil companies. The second event was the flooding of the world oil markets by the Organization of Petroleum Exporting Countries (OPEC), most notably Saudi Arabia in an effort to undermine the market share success of American oil companies that succeeded in exploiting "unconventional" shale reservoirs in many areas including North Dakota and Texas. Worldwide U.S. exports of crude oil reportedly affected OPEC's world oil exporting market share. The third factor may have involved the boycotting led by the United States world-wide against Russian produced crude oil designed to limit Russia's petroleum income due to aggression into Ukraine and other nearby emerging democracies. A similar and much earlier economic strategy that started in the early 1980's designed and implemented against the former Soviet Union was led by President Ronald Reagan and reportedly aided by OPEC's cooperation. The early 1980's effort started early in President Reagan's first term was also a world-wide oil market flood produced by Middle Eastern over production. The under reported and somewhat secret effort was designed to erode Russia's income from oil exports that was believed to be fueling their military build-up and threats to world peace. Reagan's strategy and policies appear to have been successful as the downfall of the Soviet Empire was accelerated and caused the eventual end to the Cold War and the demise of the Soviet Empire. Another well-known event was the destruction of the

Berlin Wall that produced the long-awaited unification of east and west Germany. One can argue cynically, however; that the Cold War never ended due to Russia's renewed apparent efforts to try and regain regional political and military dominance in Eastern Europe. China is also very interesting to observe as it rapidly increases its presence in Asia. One can also throw North Korea's nuclear ambitions into the charged geopolitical mix. The United States and strategic allies such as the United Kingdom and Japan once again appear to be the only major obstacles to Russia, China's and North Korea's ambitions.

Regarding the COCORP project, evidence was also obtained that indicated and possibly confirmed anticipated thrusted fault sequences in the Southern Appalachian mountain area as well as the possibility of younger sedimentary rocks at depth that have been over thrusted by older rocks. The over-thrust sub-surface fault structure concept appears and was exhibited in a memorable cross section published by the AAPG circa 1980 in a geological road map covering North Carolina and bordering states. The COCORP project endeavor demonstrated that large subsurface structural anomalies or unusual stand out features exist in the state of Georgia and other parts of the U.S. within reach of modern drilling technology. In the author's opinion, that is substantiated by examination of very limited subsurface reflection seismic survey and well log data, it appears that some of the apparent structural subsurface anomalies appear to remain untested by the drill. The opinion was inspired and indicated in the COCORP SRTSA publication referenced earlier. A critical requirement of a structural oil and gas accumulation trap is four-way closure to eliminate or at least effectively impede migration and leakage of hydrocarbons from the trap to the earth's surface. Additional seismic data coverage in the area that demonstrates four-way closure is needed over areas where structural anomalies are indicated. Stratigraphic anomalies also probably exist; however, are much harder to image with low resolution and fold (multiple redundancy sampling of the same subsurface focal point) 2D seismic data. The COCORP study serves as a model to extend structural and stratigraphic concepts to other areas especially in the southeastern U.S. The COCORP project is well documented in the literature.

Interest was further encouraged by the unique Robert B. Baum paper published in 1953 correlating the Sunniland oil field of southern Florida (near the Everglades), to the oil and gas production in southern Alabama 17 years prior to the discovery of the large Jay oil field in 1970. The Jay oil field complex is located in the panhandle of western Florida. Stratigraphy of the region was discussed with prospective correlations made into Georgia. In addition to the Sunniland oil field complex of southern Florida, the paper also discusses the Muldon, Langsdale and South Carlton fields of Alabama and the Gilbertown and Pollard fields of Mississippi. The work was quite prophetic for its time and seems to have pointed a direction to the discovery of the large Jay field of Florida. It reinforced the important role that reflection surface seismic data acquisition and interpretation would play in future hydrocarbon discoveries in the southeastern United States and industry-wide and throughout the world. Baum was a Seismograph Service Corporation (SSC) employee at the time and also presented a memorable graphic to compare the areal land mass of the state of Texas to the combined areas of the states of Mississippi, Alabama, Florida

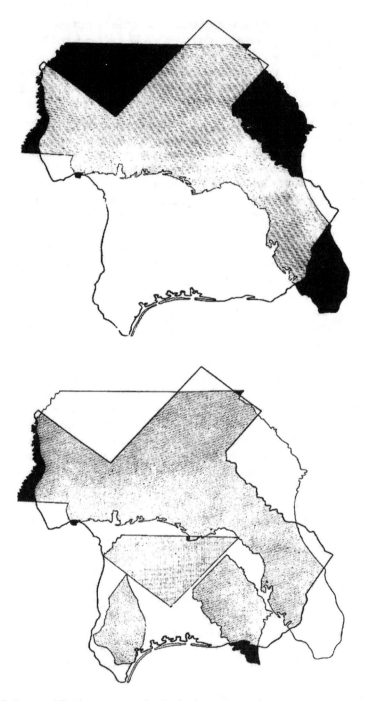

Fig. 1.2 Texas and Southeastern states land to land areas comparison
An impressive land area comparison even without including the states of SC, NC and VA

and Georgia. Baum's work is noteworthy and stands out because it is ambitious, was ahead of its time and is optimistic (Fig. 1.2).

A sizable amount of recent research has been done that includes a study of the limited literature and an examination of very limited well log data and 2D surface seismic profiles that are available within the public domain. The use of the internet, not available to earlier researchers and writers pre-mid 1990's has greatly aided timely re-examination and updates of information.

The internet or World Wide Web usage was greatly enhanced by the formation of Netscape that was the first company to attempt to capitalize on the internet's power. It was founded in 1994 and the company's first product was a practical and stable internet web browser.

Sources of proprietary data including well logs, seismic data and interpretations outside of the public domain are not included due to space and time constraints and are considered to be beyond the scope of this work. A proposal will be discussed to use modern seismic data acquisition telemetry and processing technologies and other exploration technologies including remote sensing satellite imaging such as Landsat type technologies to obtain higher resolution results to facilitate more accurate sub-surface mapping and aid in locating drilling prospects. Lower cost methods such as geo-microbial prospecting and methane detection surveys are also available. Geo-microbial prospecting is a near surface hydrocarbon exploration technique that can be used in areas of known hydrocarbon seepage such as along the Fall Line and Piedmont regions in the state of Georgia. Geo-microbial prospecting technology dates back to the late 1950's and was developed and patented by Phillips Petroleum in 1959. The survey method analyzes soil samples for the presence of bacteria growth that has been stimulated by hydrocarbon traces that have reached the near surface and surface from possible reservoirs of crude oil and/or natural gas at depth within the rocks below the surface.

I will make a distinction between surface and bore hole or down hole seismic data acquisition and processing throughout this work as they are two separate and distinct disciplines that usually do not get distinguished and discussed adequately in the bulk of the literature available pertaining to oil and gas exploration technologies. Surface seismic data acquisition has been used since the 1930's and is confined to the energy source and detectors being placed at the surface. Early borehole seismic data acquisition also dates back to the 1930's; however did not come into significant oil and gas exploration industry- wide use until the 1970's. Bore hole seismic technology, namely the vertical seismic profile (VSP) survey and the lower resolution check shot velocity survey has the geophones placed down in a well bore and may have the energy source placed at the surface, down hole or both depending on the intended use and type of data desired that may include seismic images near and away from the wellbore and governed by the total depth of the well. The seismic energy source may be a vibrator truck at the surface or an impulsive energy source such as an air gun in a water filled pit or a dynamite charge placed in a drilled shot hole. Another seismic energy source is the rock failure caused by artificially stimulating an oil and gas reservoir with hydraulic fracturing methods and observing and recording the high frequency compressional and shear wave micro seismic events using a specially designed down hole geophone tool array assembly deployed in a nearby monitor/observation well or wells a few hundred feet away. Micro seismic data may also be recorded with more difficulty and expense because of surface generated noise and land permit costs, by using geophones placed at the surface and/or in shallow drilled holes to insulate them from lower frequency surface noise. The commonly known types of borehole seismic surveys include low resolution check shot velocity surveys recorded for seismic time to depth calibrations and the more precise vertical seismic profile (VSP) surveys that yield corridor stack and/or offset seismic reflection images. Borehole seismic technology is used onshore and offshore. The most commonly used energy source for land work is a vibrator truck unit and the energy source for offshore work is a tuned air gun array powered by compressed air.

I have been involved in the data acquisition, processing and interpretation of surface seismic data since 1979 after receiving a Bachelor of Arts degree in geology and a minor in history from Bowling Green State University (BGSU) located in Bowling Green, Ohio in late 1978. The following summer I started a career in oil and gas exploration working as a seismologist and operations engineer for the Geophysical Service Inc. (GSI) division of Texas Instruments Inc. assigned to two (2) vibrator energy- sourced surface seismic crews firstly based in Pecos, Texas and later in Midland, Texas and travelled throughout the south and western U.S. including Texas, New Mexico, Montana and California involved with contracts with large independent and major oil company clients.

Texas Instruments (TI) is a large established electronics company that is credited with the early development and sales of transistor and semiconductor components world-wide. Texas Instruments was the author's second employer out of college in 1979 and was founded in 1951. TI emerged after a reorganization of its parent company GSI. GSI throughout the 1940's was a manufacturer of electronic equipment for use in the seismic industry as well as defense electronics. The GSI division had been on the market for years and was eventually sold in 1988. Without going into extraneous details; through a series of later mergers and acquisitions involving the antiquated and earlier merged GSI and Petty - Ray Geophysical (later known collectively as Geosource) assets, the remaining entities and hard capital equipment that was not scattered and sold outright, were eventually acquired by Halliburton in 1988 and then eventually sold to the now defunct Western Geophysical and then sold again and eventually absorbed by the industry giant - Schlumberger. The whole GSI and Geosource saga involving three (3) formerly leading seismic data acquisition and processing companies appeared to have finally come to an end in 2000. As a side note to make the story complete, a longtime executive of the original GSI, purchased the proprietary rights to GSI's speculative marine seismic data base in 1992 located under Canadian offshore waters, and launched a new GSI corporate entity.

The author's budding career continued at a fairly rapid pace as he was also trained in computing center and in-field seismic data processing at GSI. Later he would gain substantially more experience in surface and borehole seismic data interpretation employed as an exploration geophysicist working for the Midland,

Texas division of Texaco Inc. mapping deep structure in the Permian Basin of West Texas and New Mexico.

Texaco Inc. was a major worldwide integrated oil and gas exploration and petroleum products marketing company. It was considered part of the group of seven of the largest oil companies in the world at that time. The group was informally known as the "seven sisters", and included with additional historical details aside - Texaco, Gulf Oil, Exxon, Mobil, British Petroleum, Shell and Amoco. At present, through long and complicated merger processes, only Chevron, BP and Shell remain. Texaco Inc. merged with Getty Oil in the late 1980's and became involved in a lengthy and expensive legal battle with Pennzoil over which company had the sole right to absorb Getty Oil. Texaco Inc. was eventually acquired by Chevron in 2000. Gulf Oil had previously been acquired also by Chevron in 1985. Mobil was merged with Exxon in 1999. Amoco was merged with BP in 1998. Pennzoil after winning a huge legal battle with Texaco over Getty Oil ownership, involving Texaco's final settled payment to their much smaller rival Pennzoil of about $3 billion. The Pennzoil lawsuit contributed to Texaco filing bankruptcy in 1987 in a strategy to complicate Pennzoil's efforts. The eventual victor - Pennzoil itself later witnessed its exploration and production (E&P), assets down-sized and sold to Devon Energy in 1999. The well-known Pennzoil-Quaker State marketing entity was purchased by Shell in 2002.

The reality that consistently seems to be overlooked by the architects of many corporate merger and buy-out conundrums is the very high social and economic costs and people problems involving family finances and well-being that are created. A very large number of dedicated employees end up losing their well-paying jobs to the whims of a very few select and elite executive managers in efforts to appease Wall Street and other prominent investor demands.

The author has personally been involved in several large corporate mergers involving large global oil and gas E&P companies and service companies from 1985 to present and is of the strong opinion that there are very few true winners in these deals. The losers far outnumber the winners and include former employees and suppliers as well as the local economies that were depending on a portion the former employee's paychecks. The author knows from personal experience as a long-time oil and gas industry technical service supplier involved in sales and business development management that vendors and suppliers of the merging companies do not usually gain additional customers from such activity. That is to say, if you have 2 customers at company A and two customers at company B; hard experience shows that there is absolutely no assurance that you will in turn retain 4 customers at the merged company C. More often than not, you will be indeed fortunate enough to have 1 or 2 remaining customers at company C. In extreme cases, you end up with zero customers at company C and often have to start over from scratch. Another group put at risk with merger results are the many minority investors holding modest amounts of stock of the merging companies because their financial interests are rarely considered and can even end up being diluted and eroded. The process has gone on since well before the Industrial Revolution and shows no sign of abating and will most probably continue indefinitely in the never-ending quest for

corporate cost savings, efficiency, majority shareholder value and perceived economies of scale.

The Industrial Revolution was the transition to new manufacturing processes in
the period from about 1760 to 1840. Most of the change started in Europe and particularly England. It involved the transition from hand-made items to items made in
factories by machines. It also was the transition from an agrarian farm based workforce to a more mass production factory based workforce aided by machinery.
Company merger activity that occurred during the Industrial Revolution was rare
and the first notable mergers occurred in the United States much later after the Civil
War (1861–1865). The most notable was John D. Rockefeller (1839–1937) and his
process of growing the Standard Oil Company of Ohio rapidly in the early twentieth
century by a series of acquisitions and buy-outs of his competitors.

The author's later endeavors involved borehole seismic data acquisition and processing including VSP (vertical seismic profiling) and deep down hole and surface
micro seismic hydraulic fracture mapping (HFM) technologies for Halliburton and
later Baker Hughes working on projects located throughout North America.

Halliburton in November of 2014, formally announced that it would acquire rival
Baker Hughes Inc. in a transaction valued around $28 billion. In May of 2016, it
was announced that the Halliburton and Baker Hughes merger efforts had been
called off amid worldwide regulatory and anti-trust concerns. The world-wide oil
price crash of starting in 2015 no doubt had a detrimental effect on the merger plans.
Halliburton ended up paying Baker Hughes a $3.5 billion termination fee that was
part of the original merger agreement. As expected, the huge financial loss to
Halliburton cost many employees their jobs. Baker Hughes has since been merged
into the giant General Electric (GE) corporate giant entity. Halliburton seems to
continue to chug along as it pursues its arch-rival Schlumberger.

Since 2004, I have also been the Managing Director and owner of R.J. Brewer
and Associates, LLC involved in on-going e-commerce, technical writing and oil
and gas exploration services including seismic data acquisition, processing and
interpretation business development.

A significant amount of work over decades dating back to the early part of the
twentieth century, including subsurface mapping from sparsely drilled well control
and very limited surface seismic data acquisition has been done in the southeastern
United States. As is typical within the oil and gas exploration industry, the existence
of borehole seismic data in the area to further and more positively correlate well
bores to the surface seismic data is even more rare; however, a very few borehole
seismic surveys such as check shot velocity surveys are indicated to exist in historical well drilling compilations published by the state of Georgia. In addition, many
wells in Georgia had sonic logs run in them. The efforts were driven by the search
for oil and gas reserves extending far away from the established producing areas of
the Appalachian Basin located in New York, Pennsylvania and Ohio and the Gulf of
Mexico located in Alabama, Mississippi, Louisiana and Texas. Readily available
published information and technical work is rather obscure and dates back to the

1930's and earlier. All of the work so far has resulted in extremely limited hydrocarbon exploration drilling success except in two widely separated areas in the state of Florida. This book is a report on the attempt to compile, re-examine and re-introduce some of the notable publications that have been produced over the years in addition to introducing new evidence and conclusions to renew interest in the area. It is hoped that the recent efforts will save future workers time and money with a focus on the easily defended premise that time once spent is gone forever and therefore is the most valuable and precious resource we all possess regardless of one's youth which then rapidly becomes more evident as a person advances into middle age.

The vital work of maintaining and increasing much-needed petroleum reserves within the continental United States must continue as to produce new jobs, preserve our high standard of living and especially serve as a hedge against expensive and at times very unreliable sources of imported oil and gas as the history of twentieth century United States illustrates. This was memorably demonstrated by the Organization of Petroleum Exporting Countries (OPEC) strategy of geo-politically induced supply disruptions and trade embargos of the early 1970's.

The southeastern United States area is intriguing because it has been under explored for many years at the expense of far more promising areas such as the Gulf Coast. This is certainly understandable as the search for oil and gas is driven by stern economics as is any viable long-established business. Critics may contend that the overall geology of the southeastern United States is unfavorable for the accumulation of commercial accumulations of hydrocarbons owing to thinner sedimentary sections, ground water flushing, thermal maturation issues and scarcity of traditional source rocks and a lack of high quality seismic and well log data. The presence of the Appalachian Mountains is imposing, yet beautiful as it extends from Virginia on down to Georgia. It is intriguing why significant oil and gas production established decades ago exists on the western side of the Appalachian Mountains extending from New York, Pennsylvania, Ohio, West Virginia, Kentucky and Tennessee as part of the Appalachian Basin and into Alabama as part of the Black Warrior Basin; however, is yet to be found in even remotely sub-commercial quantities on the eastern slopes of the ancient Appalachian mountain range. Plate tectonics theory and interpreted plate suture lines in the southeastern United States will enter into the discussion. The existence of oil seeps in the state of Georgia appear to be an enigma. Unanswered questions still remain as to how to explain the occurrence and sourcing of the oil seeps in Georgia and the oil and gas shows reported in wells drilled in North Carolina, to name a few enduring puzzles. One theory may be that the oil is sourced from untapped reservoirs offshore and has migrated over the eons many miles eastward onshore slowly reaching the earth's surface. We need to be reminded that only the drill finds the oil and gas reservoir. Additional work in the area is also needed to test new and established theories and learn more about the complex geologic history of a fascinating and extensive province of the continental United States.

Bibliography

Baum RB (1953) Oil and Gas Exploration in Alabama, Georgia and Florida. AAPG,Tulsa, pp 340–359

Cook FA, Brown Larry D, Kaufman S, Oliver JE (1983) The Cocorp Seismic Reflection Traverse Across the Southern Appalachians, AAPG Studies in Geology No. 14. The American Association of Petroleum Geologists, Tulsa p 61

Chapter 2
Geology of the Southeastern United States

Thoughts of the area of the southeastern United States may conjure up images of a vast coastal plain with sediments of sand and silt and lowlands of rich farmland including cotton fields of the antebellum South and the tide lands with occasional swamplands and meandering large rivers increasing in elevation northwestward up to the Fall Line entering extensive evergreen and deciduous forests and nearing the ancient and heavily-eroded Appalachian Mountain front.

The Fall Line is a prominent northeast to southwest trending topographic feature characterized by an abrupt change in elevation as one travels down from the higher Piedmont provinces to the lower Coastal plain provinces present in the southern states of Alabama, Georgia, South Carolina, North Carolina and Virginia and extending even farther north into Maryland, Pennsylvania, New Jersey, New York and Connecticut. In other words, the Fall Line is the physiographic separation trace between the Coastal Plain and Piedmont provinces of the east coast of the United States (Fig. 2.1). The Fall Line in the state of Georgia is a prominent northeast to southwest trending topographic feature. The cities of Columbus, Macon, Milledgeville and Augusta are all located near the Fall Line.

Large coastal plain areas of the states mentioned above lie between their fall lines and the Atlantic Ocean. All of the vast coastal plain area in the southeastern United States is very under-explored with a lack of deep well drilling and all continue to be prospective regarding possible commercial oil and/or gas accumulations.

Historically, steamboats and other vessels serving the cotton plantations, farms and other industries of the area could travel no farther inland and upstream plying the larger and navigable streams such as the Tombigbee River in Alabama and Mississippi, the Flint River in Florida and Georgia, the Santee River in South Carolina, the Neuse River in North Carolina as well as the James and Potomac Rivers in Virginia and the Chesapeake Bay waterway in Virginia and Maryland. Important waterways even farther north are Delaware Bay and the Delaware River between Pennsylvania and New Jersey and the Hudson River between New York,

R. J. Brewer, *Hydrocarbon Potential in Southeastern United States*,
SpringerBriefs in Earth Sciences, https://doi.org/10.1007/978-3-030-00218-3_2

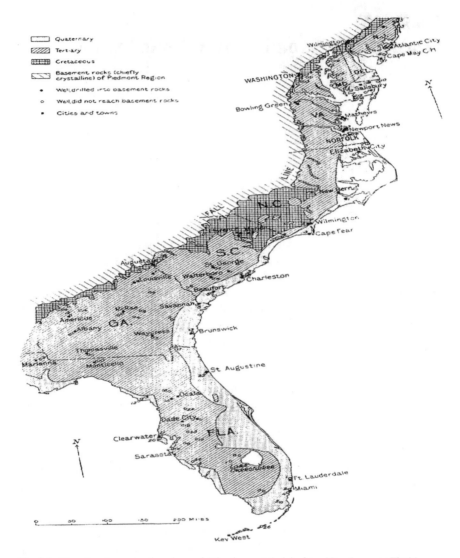

Fig. 2.1 Map showing general geology of Atlantic coastal plain from New Jersey to Florida
The Fall Line is indicated by dashed lines

Connecticut and Massachusetts. The boats could travel no farther inland and up some incised eroded out river valleys because they would have encountered water-falls and rapids present as the streams flow down from the Piedmont and Appalachian highlands on their way to the Atlantic Ocean. This fact helped stimulate the devel-opment of railroads in the region.

The Appalachian Mountains of the eastern United States started to form during the Pennsylvanian period of the Paleozoic era about 135 million years ago.

At first appearance, a petroleum geologist may consider the area a prime region for the exploration of oil and gas much like the petroleum rich U.S. Gulf Coast of Louisiana and Texas and to a lesser extent – the Florida panhandle, Alabama and Mississippi. Indeed, this was probably what many petroleum explorationists including geologists and geophysicists of the first half of the twentieth century envisioned. The first half of the twentieth century experienced a fair amount of drilling with very limited success. The state of Georgia reportedly in the 1950's initiated an offer of a cash prize that eventually reached $one million to award the first oil company to establish an oil well capable of producing 200 barrels a day in the state. The cash bounty was then lowered to $250,000 commensurate with the discouraging lack of drilling success in Georgia. The state of Georgia confirmed in 2017 that the cash prize is still in force.

The southeast United States and Atlantic coastal basin area is believed to be the only large sedimentary basin in the United States that has documented oil seeps such as in the state of Georgia that have not been confirmed by well drilling resulting in commercial oil and gas field discoveries. In other large basins such as the Appalachian basin in Pennsylvania, Ohio and Tennessee, established oil seeps first observed hundreds of years ago by early settlers in the seventeenth and eighteenth centuries seeking brine and salt have pointed the way to later successful wildcat drilling and the discovery of highly commercial quantities of oil and gas. The petroleum discoveries in Ohio, Pennsylvania and New York greatly aided the fortunes and growth of John D. Rockefeller's empire and his Standard Oil Company of Ohio (Sohio) and the Standard Oil Company of New York (Socony) later known as Mobil as well as the Standard Oil Company of New Jersey much later known as Exxon.,

Large oil seeps in California, the most notable being the famous La Brea Tar Pits located in Los Angeles County pointed the way to large hydrocarbon accumulations in the area. The seeps are located in a large petroleum producing area known as the San Joaquin Basin. The La Brea Tar Pits are composed of asphalt and bitumen residue that is left behind after the more volatile components of the oil evaporates into the atmosphere. The Signal Peak, Long Beach and Huntington Beach oilfields are but a few of the more famous oilfields that are within and surrounded by the second largest city in the United States of Los Angeles. The La Brea Tar pits are sourced by porous and permeable Tertiary age rock formations that contain large accumulations of oil and gas thousands of feet down dip (downhill) underground from the La Brea Tar Pits. The hydrocarbons migrate up moving from the deep Tertiary lower rock formations and then enter the younger Pleistocene age beds before eventually reaching the surface to become oil springs. Most of the oil reaching the surface evaporates and escapes into the atmosphere along with the associated natural gas to leave behind tar, asphalt and bitumen. The oil seepage process has been going on for many thousands of years.

A northwest to southeast conceptualized geologic cross section drawn through the La Brea Tar Pit area indicates that the oil is slowly leaking up from Tertiary age porous and permeable rock formation reservoirs at depth and located down dip from the oil seeps that come to the surface and collect in low lying valley areas. A large fault system has also been identified that cuts through most of the Tertiary rock formations and facilitates oil seepage by acting as a conduit up into the overlying Pleistocene age sediments and valley fill. The oil finally reaches the surface to collect as pools in low lying areas and valleys (Fig. 2.2).

The oil and gas seeps in California reaching the surface are numerous and form a large and wide northwest to southeast trend somewhat centered along the coastline. The seeps are up dip from large producing oil and gas fields located deep underground. The producing fields also exhibit a predominately northwest to southeast trend and are located miles to the east of the seeps.

Early settlers considered the oil seeps in the eastern United States and Appalachian Basin and probably in California as well as a nuisance that at times tainted their water wells and got in the way of their salt gathering efforts. The oil was reportedly separated from the brine and salt by soaking it up with blankets. Some oil was gathered however and used as a lubricant and as an ingredient in liniments and medicines. Later in the nineteenth century, chemists figured out that crude oil could be refined via the hazardous process of fractional distillation and refined into commercial amounts of kerosene and other similar distillates such as naphtha and used for cleaning, illumination, cooking and heating.

Gasoline refined from crude oil was an even more hazardous endeavor that was not widely practiced until after Henry Ford's improvements in the carburation of the internal combustion gasoline engine and legendary mass production technology involving the development, production and marketing of reliable automobiles in the

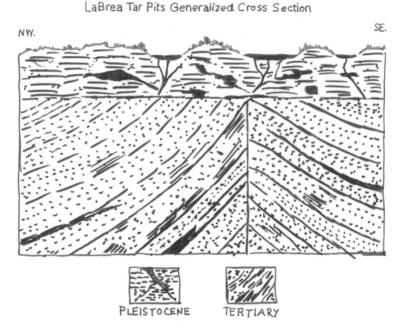

Fig. 2.2 Generalized NW-SE cross section of the La Brea Tar Pits of Southern California Oil accumulations are noted by solid black areas

early twentieth century. The presence of gasoline mixed with kerosene was reportedly discovered by accident when refiners tried to increase yields of kerosene from crude oil and ended up including the much more volatile, flammable and at times explosive higher distillate later to become more widely known as gasoline in their production of kerosene.

Oil seeps and tar pits in southern California appear to have a similar history eventually leading to the discovery of substantial and very commercial accumulations of hydrocarbons in the state and helped fuel the growth of companies such as the Union Oil Company of California (Unocal) and the Standard Oil Company of California, later known as Chevron.

Tracing surface petroleum seeps has proven to be a reliable method of exploration world-wide in areas such as Mexico and the Middle East along with the application of the then anticlinal theory of hydrocarbon accumulation which has become well established in many hydrocarbon producing areas and is widely believed to be a scientific fact.

There are two theories regarding the origin of hydrocarbons. The theory of organic hydrocarbon generation and the inorganic or abiotic theory of hydrocarbon generation. The first and most accepted and established is the organic theory of hydrocarbon generation involving the decomposition of organic matter at depth that has collected in a sedimentary rock. The most common and widespread sedimentary rock is shale. Shale is lithified or turned to rock mud that has been buried and then compressed by the weight of the rocks above over the time of millions of years. The organic matter in the shale is widely believed to be the buried remains of ancient plants and animals that has decomposed into oil and natural gas. These hydrocarbons then found their way and migrated up into more porous and permeable rock such as sandstone, limestone and dolomite. The organic decomposition process is sometimes observed at the surface in swamps and peat bogs that produce small quantities of methane and even oil. The generation of methane has also been observed from the decomposition of garbage at landfills. Methane (CH_4) is a flammable gas, occurs widely in nature and is the main component of natural gas.

The inorganic or abiotic theory of hydrocarbon origination is far less favored by scientists in recent times and has to do with hydrocarbons – primarily methane gas being generated far below the earth's surface near the mantle and core and eventually migrating up toward the surface to become trapped and accumulate underground or escape onto the land surface or seabed.

The inorganic theory can serve to help explain the very large oil and gas reservoirs found in the deep water (over 4000 ft. water depth) discovered in the Gulf of Mexico and other deep marine locations of the world. The formation of hydrates that may contain methane gas is sometimes encountered at the seafloor in deeper waters. The origination of hydrates is not easily explained by the organic theory of hydrocarbon formation and seems to be more easily explained by the abiotic theory. Reinforcing the abiotic theory is the idea that a very large portion of the ocean areas thousands of feet below the ocean's floor now producing hydrocarbons, were always under water and never exposed at the surface to facilitate enough animal and plant remains to first exist, thrive and then die and get buried as indicated by the organic theory throughout geologic time. The inorganic theory has been used to explain the origin of oil and gas fields found in very remote areas such as Kazakhstan where hydrocarbons were not thought to exist before due to previous notions that there was an absence of source and reservoir rocks.

It is interesting that both the organic and inorganic theories of hydrocarbon generation were reportedly first hypothesized and documented by Russian geoscientists back in the first half of the twentieth century.

It is very intriguing to the author that the very modest oil seeps observed in the East Coastal basin specifically along the piedmont province located in the central part of the state of Georgia and trending along the Fall Line southeast of the Appalachian Mountains may be the only oil seep area in North America that has not pointed the way to larger accumulations of hydrocarbons at depth. Many drilling attempts have been made since the early part of the twentieth century, yet no oil or gas fields have been discovered and situation remains an enigma. The source and subsurface reservoir for the oil seeps remains a mystery. Perhaps the abiotic theory of hydrocarbon origination may help explain and indicate that the oil seeps originated from great depth and migrated to the surface via deep faulting and fractures and passed through crystalline basement rocks and accumulated under Precambrian age igneous rocks and are trapped in younger Ordovician age sedimentary rocks that were thrusted over by the older rocks. The search will continue at a slow pace until definitive answers are obtained. Only the drill bit finds the oil and gas.

One of the largest discoveries of oil in North America traceable from the long-time observation of surface oil seeps as substantial evidence of large accumulations underground is the Golden Lane trend located between the cities of Tampico and Tuxpan, in the state of Veracruz, Mexico. A large portion of the Golden Lane trend was mapped by an exploration team led by the brilliant and renowned geologist Everett DeGolyer (1886–1956) and others during the early part of the twentieth century (1909–1914) before World War I and during the Mexican Revolution. DeGolyer in addition to being a gifted geoscientist, was a visionary who foresaw the vast and as yet undiscovered petroleum potential of Mexico. He was employed by the English oil company El Aguila and did extensive field geology on foot and from horseback and mapped buried hills or anticlines on several large ranches (haciendas) in an area that had few outcrops of rocks to help discern the subsurface. DeGolyer used dry stream beds (arroyos) and volcanic plugs exposed at the surface as large hills (cerros) to find exposures of rock formations – most notably the Tamasopo (El Abra) carbonate (limestone) to make dip and strike measurements using a Brunton compass and plane table alidade for points of control or elevations. The control points were later used back at DeGolyer's campsites in the evening after long days out in the field to make contour maps to locate anticlines that he believed would act as petroleum traps in the subsurface. The Golden Lane Trend is a series of buried predominately rustid (snail-like marine organisms) reefs lined up in a circular (atoll) arrangement that produces hydrocarbons primarily from the Cretaceous Tamasopo (El Abra) carbonate (primarily limestone) formation. Potrero del Llano is a notable oil field in the Golden Lane trend and was the first commercial field discovered by DeGolyer and his team.

On December 27, 1910, the Potrero # 4 well struck oil at 1912 feet with initial flow rates estimated at over 100,000 barrels (bbls.) per day that soon made newspaper headlines world-wide. Adequate oil storage facilities did not exist at the drill site and the well flowed out of control for many days as armies of workers toiled day and night to shore up the vast amounts of oil at the surface building catch ponds and diverting the oil in streams and then burning large portions of it off at safe distances downstream. Fortunately, the well did not catch fire with a lightning strike or other mishap that tended to plaque some early oil discoveries after wells blew out in the Appalachian Basin of the northeastern United States such as the area around the newly-discovered Lima field in northwestern Ohio in 1907.

The Potrero # 4 discovery well blow out in Mexico was clearly an environmental disaster by any standards and took over 3 months to contain and clean up to then acceptable levels. Protection of the environment, ecosystems and sustainable development concepts, appreciation and practices were virtually unknown at the time. The well's enormous flow from virgin reservoir pressure was eventually choked back to about 45,000 bbls. Per day in the spring of 1911. The well flowed a huge amount of oil for a single borehole in a short period of time. Crude oil was selling for about $17 per bbl. at the time.

Significant oil seeps had been observed in the coastal areas of the southern Gulf of Mexico for centuries and pointed the way for the discovery of many oil fields in east central Mexico including Furbero, Tanhuijo and Potrero del Llano. DeGolyer's theories were proven to be correct by drilling results and history indicates that he appeared to have succeeded beyond his wildest dreams. The discovery of the Golden Lane oil fields thrust Mexico into the role of a leading oil exporter in a very short period of time and caught the attention of the governments and navies of the United States, Great Britain, France, Germany and Japan who quickly opened consulates in Tampico as they competed for reliable sources of crude oil to fuel their warships that had followed Sir Winston Churchill (1874–1965) of the Royal Navy and Great Britain's lead and been converted from coal to fuel oil. Fuel oil has significant storage and heating advantages over coal.

Reportedly, one (1) gallon of residual fuel oil refined from crude oil has a value of 136,000 British Thermal Units (BTU). One (1) pound of coal depending on the specific type of coal, has a value of only approximately 12,500 BTU. By comparison, one (1) cubic foot of natural gas has a value of approximately 1075 BTU. The comparisons underscore why crude oil has commanded high prices and demand compared to other fuel types over many decades and especially during wartime. The continued high demand and usage of crude oil refined products and natural gas is predicted to endure for many years to come until science and technology can someday again produce suitable replacement(s) that meet the world's high BTU requirements for affordable and readily obtained energy.

In some residential areas of the United States and world-wide that are too far from municipal natural gas transmission lines, people must rely on bottled gas or LPG (liquified petroleum gas) such as propane and/or butane for their heating needs, if available. Although significantly more expensive than natural gas – one (1) cubic foot of propane or butane is equivalent to 2570 BTU and 3260 BTU respectively.

If one accepts the premise that most energy on earth can be eventually traced back to the sun and as by - products of solar energy – it would seem that the harnessing of solar energy should continue to warrant vigorous research and development. Solar energy lends itself to being converted to electricity quite well; however, battery storage technology seems to still be advancing at a snail's pace. Nuclear energy is a proven option and intriguing; however still awaits substantial future technological advancement and to ease widespread safety fears. The issue of what to safely do with and dispose of the radioactive waste fuel from nuclear reactors promises to continue to be a vexing problem with no easy solution. Nuclear energy has a problematic past and public image world-wide. The radioactive fuel required for nuclear power plants whether it be new fuel or spent fuel has the potential to be misused and perhaps concentrated and then refined and possibly weaponized by troublemakers.

Crude oil and natural gas will continue to dominant and be the best choices of energy for the world until a more cost-effective alternative can be found that can also deliver the portability and concentration needed to be easily converted to heat and to reliably power our transportation and electricity generation needs on a global scale.

The petroleum exploration methods used by Everette DeGolyer, although perhaps somewhat basic by modern standards can be applied again in the Southeastern United States today. Basic field geology practices are often overlooked today in favor of higher technology and costlier methods such as reflection seismic profiling. Interestingly, DeGolyer was an early promoter and user of the then new technology of reflection seismic profiling that was first proven in his home state of Oklahoma by John Clarence Karcher who applied for a patent in 1919 and confirmed the technology with his well-known and documented seismic data field trials in 1921.

A somewhat similar mapping effort to the very successful methods used by DeGolyer in Mexico was undertaken in Georgia in 1917 by a team of state geologists led by S.W. McCallie to try and map an anticline structure in the near-surface in the vicinity around the famous oil seeps in Telfair County, Georgia. The report published by McCallie in 1919 is the best scientific account of the famous oil seeps in the state of Georgia that the author could find to date. Field geology was done along river and road access along the Piedmont and Fall Line areas including the innovative and unconventional idea of measuring the water levels in some of the local residents' water wells to add more points of control and hopefully help indicate four-way rock dip closure that might point the way to an anticlinal oil trap within a few thousand feet of the surface. Unfortunately, conclusive evidence of an anticline was reportedly not found during the mapping project. Even so, the recommendation was made to drill a test well to at least the crystalline basement rock estimated to be encountered at a depth of about 3000 feet. McCallie was an optimist and also stated in his report that he believed oil reservoirs to exist in Georgia in addition to indicating that the oil seeps were live oil.

The coastal plains of the Southeastern United States are dominated by Cretaceous age sediments. Formations such as the Wilcox, Tuscaloosa, Eutaw and Ripley or equivalents are identified that are known to be productive in well-known areas of the Gulf Coast geosyncline to the west in Alabama, Mississippi and Louisiana (Figs. 2.3, 2.4, 2.5, 2.6 and 2.7).

Fig. 2.3 Map of Southeastern states showing cross sections A–B and C–D
The 1953 map pre-dates the 1970 discovery of the large Jay oil field complex in the panhandle of
Florida by 17 years. Map also pre-dates later discoveries and increased hydrocarbon production in
Black Warrior Basin of MS and AL

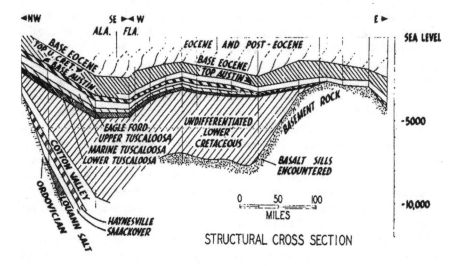

Fig. 2.4 Alabama and Georgia east to west A–B generalized cross section

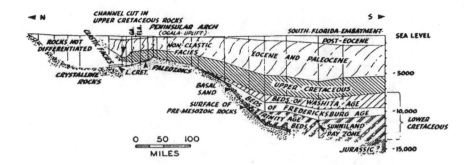

Fig. 2.5 Generalized north-south cross section C–D from Georgia through Florida

Most of the wells drilled in Georgia were TD'd (total depth) after drilling only a few thousand feet and usually at the base of Cretaceous age sediments soon after "basement" crystalline igneous rocks were tagged by the drill bit. Most of the petroleum exploration wells drilled in Georgia reached down to the 4000–6000 ft. depth range. Drilling deeper could become costlier and time consuming because of the much harder igneous rock that would have to be penetrated. A few more ambitious operators than the average prospectors with deeper pockets and financial resources drilled farther down with more innovative ideas in hopes of penetrating through hard crystalline basement sequences and encountering softer and prospective sedimentary rock underneath. Drilling records and wire line logging data from such exploits are not readily available in the public domain and will require further research to get answers. The relentless passage of ideas and concepts into the mist of time makes evidence harder to obtain and substantiate that may indicate the presence of hydrocarbons from dated drilling efforts. Oil and gas exploration has always been a very competitive business. Often, well test data is proprietary to the well operator and a closely guarded secret with usually only minimal amounts of data released per law to state agencies and geologic surveys. Significant oil and gas discoveries become public knowledge rather quickly however; there-fore the general consensus in the oil and gas industry is that the Southeastern United States is not a very prospective area for hydrocarbons. Yet the intrigue remains because of the presence of some stubborn unexplainable oil seeps in central Georgia and documented oil and gas shows in some wells in North Carolina and commercial oil and gas production in eastern and southern Florida along with a few scattered oil and gas shows in wells in between. One conclusion is that not enough wells have been drilled in the Southeastern United States and that it is only a matter of time, investment, effort and statistics before a significant new discovery is made. On a worldwide basis, new significant oil and gas discoveries happen every few years. Why should the Southeastern United States with hundreds of thousands of square miles of area including onshore and offshore be an exception?

The sedimentary rock sequence in the southeastern United States compared to the Gulf Coast is relatively thin above the crystalline Pre-Cambrian basement rock. Most of the exploration work in the region has been done in the state of Georgia and

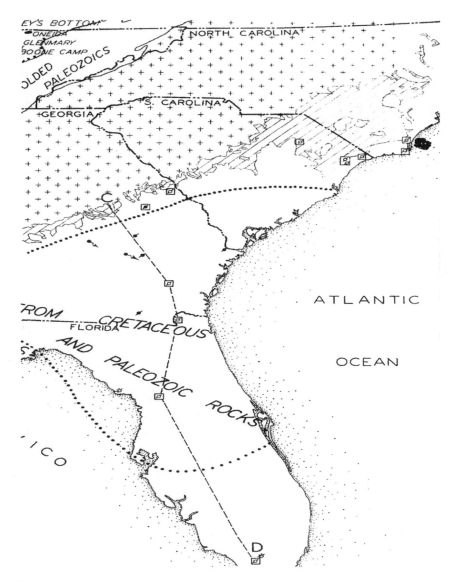

Fig. 2.6 Map showing northwest to southeast (C–D) cross section from Georgia through Florida

many wells drilled in Georgia penetrated a few thousand feet of sediment before reaching total depth (TD) to be completed in hard and tight igneous rock. Igneous rock was formed by the solidification of rock from a mobile and hot molten or partially molten state. Rocks of igneous origin may on rare occasion serve as hydrocarbon reservoirs if they are heavily fractured from tectonic activity such as faulting resulting in significant porosity and permeability. Some formations of igneous origin such as the Granite Wash in the Texas and Oklahoma panhandle which

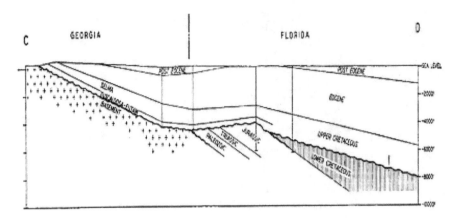

Fig. 2.7   Generalized north-south cross section C–D from Georgia through Florida

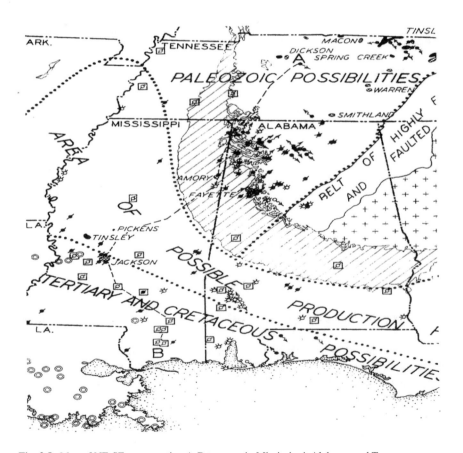

Fig. 2.8   Map of NE-SE cross section A-B traverse in Mississippi, Alabama and Tennessee

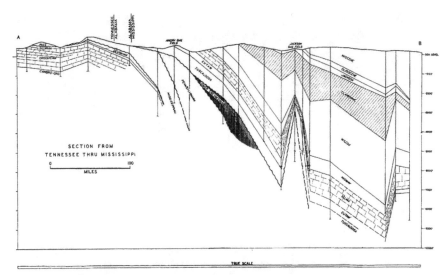

FIG. 3.—Generalized structure section *AB* from Tennessee through Mississippi. Line of section shown in Figure 1.

Fig. 2.8a Generalized structure section A-B from central Tennessee through southern Mississippi. Demonstrates hydrocarbon production structural features from Gulf of Mexico coastal plain of Mississippi north to western slope of southern Appalachian mountain complex in Tennessee. Provides model for similar structural features in southeastern U.S. along coastal plain from eastern Florida all along east coast as far north as up and into Maryland and beyond paralleling fall line

are actually conglomerations of predominately broken-up igneous rocks eroded from the ancient and partially buried Arbuckle Mountain range have proven to be good reservoirs and produce significant amounts of hydrocarbons. Notable wells in the Southeastern United States exist that were drilled much deeper through Cretaceous age sediments indicating that the total sedimentary rock section is actually much thicker than first believed. The presence of sedimentary rock perhaps capable of bearing hydrocarbons has been postulated to exist below igneous rock previously thought to be the basement limit to marine sedimentation. The theories involving the existence of sediments below the basement rock was probably further illuminated by deep drilling results in Georgia in the early 1980's in a deep test conducted by Chevron. One of the deepest wells drilled in Georgia was the A.P. Snipes et al. No. 1 in Jeff Davis County completed in June 29, 1981 by Chevron U.S.A. Inc. to a TD of 11,470 ft. Like all other wells in Georgia drilled in the search for hydrocarbons, the Snipes location was a dry hole. Further research is required to determine whether any shows of oil and gas were encountered along with the detailed lithology of the rocks penetrated. Reportedly, hydrocarbon shows were encountered in the Chevron well according to the author's personal sources who are geological and geophysical personnel familiar with oil and gas exploration in Georgia conducted in the late twentieth century. Chevron presumably had a significant oil and gas prospect mapped from a thorough mapping of acquired proprietary surface seismic data and well control and perhaps surface geology to justify to man-

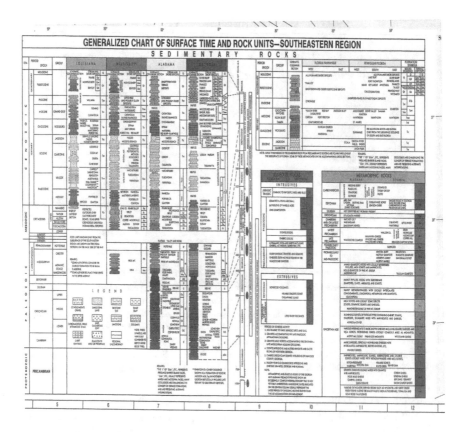

Fig. 2.9 Stratigraphy of the southeastern United States

agement the large expense of drilling a deep rank wildcat hundreds of miles from any established production. The deep Chevron test is located near Scotland, Georgia in nearby Telfair County where small oil and gas seeps have been observed on the old Samples farm property for decades.

A copy of a suite of electric wire line logs was obtained and examined by the author from the state of Georgia for the Chevron A.P. Snipes, et al. location in 2017. The 11,470 ft. well appears to have TD'd deep in the crystalline basement underlying the coastal plain like nearly all the other oil and gas exploration dry holes in Georgia, yet some anomalies appear on the sonic and density logs examined. The log anomalies may indicate the presence softer sedimentary rock sandwiched in between the igneous rock from 7500 ft. to TD. The operator may have been trying to test the theory that sedimentary rock had been thrusted in below the igneous country rock.

Other evidence of sediments existing beneath previously established basement rock in Georgia is indicated from reflection 2D seismic data gathered during the COCORP project of the early 1980's and perhaps further supported by even earlier gravity and magnetic survey work.

Sedimentary sequences in Florida, South Carolina, North Carolina and Virginia are of course different and less is known about the subsurface in these states because few hydrocarbon exploration wells have been drilled. The exceptions are eastern and southern Florida and western Virginia where commercial oil and gas production exists and will serve as models to encourage more exploratory drilling in the areas beyond (Figs. 2.8, 2.8a and 2.9).

Bibliography

Pfiester SL (2011) The Golden Lane (Faja de Oro). Chengalera Press, Sam L. Pfiester, p 277
Postley OC (1938) Oil and gas possibilities in Atlantic Coastal plain from New Jersey to Florida. AAPG Bull 22(7):799–815

Chapter 3
Oil and Gas Production in the United States

After studying large oil and gas production maps of the entire United States for a while, certain patterns may start to emerge. It is obvious where the established oil and gas production is and where it is not. After more viewing, it may seem that oil and gas production trends are seen and the gaps between them result in questions as to why there is oil and gas found in some areas and not in others. Fig. 1.1. The question as to why commercial accumulations of hydrocarbons exist in some areas and not in others is due to geologic and stratigraphic reasons. The reasons involve subsurface rock deformation and structure and the specific lithology and petrophysics of the possible reservoir quality rocks having favorable porosity and permeability encountered at depth. More wells will need to be drilled in non-producing areas to further test the current ideas. On the maps, oil production is usually represented in red, gas production is in green and dry holes are in gray.

Specifically relating to the southeastern United States, a limited amount of production is seen in south Florida involving the Sunniland subsurface reef feature and in the panhandle or far western part of Florida near Alabama. Most of the production in the panhandle of Florida involves the large Jay field subsurface structural feature the majority of which is located in Florida and extends up into Alabama. The Sunniland and Jay field complexes, although hundreds of miles apart are believed to be part of the huge and extensive Gulf Coast geosyncline that also involves the significant oil and gas reserves of the Gulf Coasts of Alabama, Mississippi, Louisiana and Texas. The Gulf Coast geosyncline is also underneath a substantial amount of area offshore of all the states mentioned.

At present, it appears that the only substantial evidence of hydrocarbon production anywhere within the East Coastal Basin was discovered in the 1980's many hundreds of miles to the northeast of offshore New Jersey in the Baltimore Canyon area. At the time, the mostly natural gas accumulations were deemed to be non-commercial.

© The Author(s), under exclusive licence to Springer Nature Switzerland AG 2019
R. J. Brewer, *Hydrocarbon Potential in Southeastern United States*,
SpringerBriefs in Earth Sciences, https://doi.org/10.1007/978-3-030-00218-3_3

Focusing one's attention even much farther to the northeast in Canadian waters, significant amounts of natural gas were discovered in the Sable Island area. The Sable Island natural gas production is now a natural gas field complex composed of several separate gas fields. The Sable Island production may be too far northeast to be considered part of the extensive and elongate East Coastal Basin. More research and data examination will be needed to more clearly indicate if the Sable Island hydrocarbon production system is part of and a northeastern extension of the elongate and extensive East Coastal Basin or a separate sedimentary basin feature.

The Sable Offshore Energy Project (SOEP) is a consortium based in Halifax, Nova Scotia which explores for and produces natural gas near Sable Island on the edge of the Nova Scotian continental shelf in eastern Canada. SOEP produces between 400 and 500 million cubic feet (14,000,000 m^3) of natural gas and 20,000 barrels (3200 m^3) of natural gas liquids every day.

The first tier was completed in 1999 resulting in the development of the Thebaud, North Triumph, and Venture natural gas fields

In 1979 the drill rig Gulf Tide struck success and the Venture field was discovered. At the time, the cost of developing the gas field proved prohibitive. Extreme North Atlantic Ocean weather and low natural gas prices were blamed. In 1995, improvements in drilling technology along with an increase in natural gas prices made recovering the gas economic allowing SOEP to emerge.

The Venture Gas Field was discovered in 1979 when the Venture D-23 well was drilled and is located a few miles of the east end of Sable Island on the Scotian Shelf in 12–20 m of water.

The structural trap consists of an east-west trending anticline bounded on the north by a fault, with gas occurring in multiple Upper Jurassic to Lower Cretaceous sandstone reservoirs 1600 m thick at depths from 4 to 6 km.

Back to the southeastern United States, oil and gas fields remain to be discovered in Georgia, South Carolina, North Carolina and most of Virginia except for the extreme western part of Virginia near West Virginia. Unsuccessful drilling is evident in Georgia and the Carolinas by numerous abandoned exploration well locations. What is not clearly evident from the map is the limited drilling that has occurred offshore of these areas. The offshore areas also continue to be prospective for oil and gas accumulations; however, involve substantially higher operating and leasing costs and environmental concerns and constraints.

Bibliography

American Geological Institute (1976) Dictionary of geological terms. Anchor Press, Revised Edition, p 221
Hull JPD, Teas LP (1919) A preliminary report on the oil prospect near Scotland Telfair County. Georgia, Geological Survey of Georgia, Index Printing Company, Atlanta, GA, 31 pages
Spencer JA, Camp MJ (2008) Ohio oil and gas. Arcadia Publishing, p 127
Steele WM (1986) Petroleum exploration wells in Georgia 1979–1984, # 77. Georgia Department of Natural Resources

Chapter 4
Triassic Rift Basins in the Southeastern United States

Convincing evidence of the existence of Triassic age rift or extension stretch basins is found in Kentucky, Pennsylvania, West Virginia, Virginia, North Carolina, South Carolina and Georgia. The most notable rift basin feature is the Rome Trough that extends from eastern Pennsylvania southwest through West Virginia and into eastern Kentucky. Indications of extensions of the Rome Trough from basement faulting may be that the feature even reaches north to southern Ohio and south into Tennessee. The Rome Trough has been extensively drilled in Pennsylvania and West Virginia and the western edge in southwest Pennsylvania corresponds to highly naturally fractured areas with high oil and natural gas production rates of the very productive Marcellus shale play currently being exploited with the latest horizontal drilling and hydraulic fracturing technologies.

The Triassic basins are surrounded by areas of predominately igneous and metamorphic rocks. Within the basins have been found red colored formations known as "red beds" of sandstone, conglomerate and silt that were down faulted into the older metamorphic and igneous formations. The "red bed" formations continue to be explored as potential hydrocarbon reservoir material in spite of the presence of igneous dikes and sills that intruded into the sediments and may have destroyed some oil and gas accumulations.

The principal Triassic basins include the Newark Basin extending from the Hudson River southwest to the Schuylkill River. The Gettysburg Basin which extends from the Schuylkill River to Maryland. The Culpepper Basin that extends from the Potomac River south to Culpepper, Virginia. The Richmond Basin that is located a few miles west of Richmond, Virginia. The Danville Basin of southern Virginia and the Deep River Basin of North Carolina. The northern most Triassic basin in the United States may be the Hartford Basin of New England that extends along the Connecticut River from New Haven northward almost to the state line of Massachusetts and Vermont.

Some parts of the Triassic rift basins in Virginia, North Carolina, South Carolina and Georgia have also been penetrated by wells searching for hydrocarbons with no

© The Author(s), under exclusive licence to Springer Nature Switzerland AG 2019   31
R. J. Brewer, *Hydrocarbon Potential in Southeastern United States*,
SpringerBriefs in Earth Sciences, https://doi.org/10.1007/978-3-030-00218-3_4

success up to time of this writing. Not much is known about the geology, structure and stratigraphy of the Triassic rift basins especially in North and South Carolina because of very sparse deep drilling by operators. The drilling data obtained is usually proprietary and runs the risk of becoming lost and forgotten as time quickly passes. The concern is further complicated by the volatile economics and nature of the oil and gas exploration business especially in the United States in recent years since late 2013 (Fig. 4.1).

The Triassic period of the Mesozoic era on the geologic time scale is generally believed to have spanned 35 million years approximately 190 to 225 million years ago. The Triassic period was the first period of the Mesozoic era after the end of the extensively long Paleozoic era that lasted some 345 million years starting with the Cambrian period and ending with the Permian period. Animal life forms during the Triassic period as evidenced from the fossil record contained in Triassic age dated rocks include theodont and therapsid reptiles and seed plants known as gymnosperms that include conifers and ferns. Some flowering plants may have also existed as early as the Triassic. References to the lithology of the Triassic in the southeastern

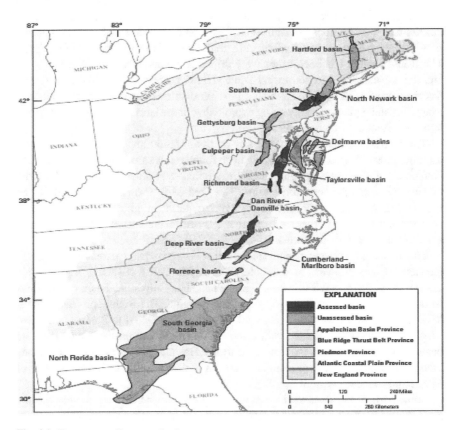

Fig. 4.1 East coast sedimentary basins
The map also indicates the on-shore portion of the east coastal basin

United States sometimes mention "red beds" alluding to the red color of some of the rocks observed and mapped in outcrop exposed at the earth's surface over the years. Per plate tectonic theory, Triassic rift basins were believed to have been formed in the southeastern United States when the American plate separated from the African plate. Both plates were formerly part of the super continent known as Pangaea during the Paleozoic era. Rift or basin depressions were created by the separation that eventually filled with Triassic age sediments.

During the Triassic period, the dominant theory is that a supercontinent known as Pangaea began tearing apart and rifts developed between North America and the African portion of a large landmass known as Gondwanaland. As the earth's crust stretched, large blocks slowly moved down creating physiographic basins in which thick beds of red- stained sandstone, siltstone and shale were deposited. Today, some of these rock sequences are known as "Triassic red beds." The sedimentary rocks were intruded by sills of dark igneous basalt such as the one exposed in New Jersey's Palisades along the west bank of the Hudson River. Narrow elongated remnants of the Triassic basins also occur from Virginia north to Nova Scotia. Rift valleys and volcanism occurred during the Triassic in the eastern United States and Canada about 225 million years ago. During the Jurassic, the earliest evidence of continental drift and volcanism in the western United States is believed to have occurred about 175 million years ago.

Fossils found in Triassic age rocks indicate that the general climate was warm during that long span of time. During the Triassic, much of North America has been interpreted to lay between the equator and 30° north latitude so that subtropical conditions existed as far north as Wyoming and New England. The terrain was dominated by evergreen trees, most of them conifers and ginkgoes. The cycad palms and scale trees that were the most prominent flora of the preceding Permian period are believed to have existed, however were not as numerous or varied as the evergreens.

The most significant animals to appear in the Triassic were the much-storied dinosaurs. The earliest members of this group were not nearly as large and impressive as the huge dinosaurs of the later Mesozoic and were apparently no more than 10–15 feet long as evidenced by their fossilized remains. Other well- known reptiles of the Triassic were the long snouted ichthyosaurs. These carnivores apparently had impressive sleek bodies that resembled modern dolphins and porpoises and the distinctive carnivorous plesiosaurs with their broad turtle-like bodies, long necks and tails and large flippers. Some legends suggest that descendants of plesiosaurs may still exist in the oceans and even in fresh water. One hypothesis regarding reported sightings of the mysterious Loch Ness monster over centuries is the idea that it may be a descendant of a plesiosaur-type creature. A creature somewhat resembling a plesiosaur was reportedly netted by Japanese fisherman about 20 years ago and was later discounted as being a kind of large odd-looking shark. There are also reports that plesiosaur like animals have been sighted near the shores of Lake Erie – the shallowest, warmest and most southern of the Great Lakes of North America. A famous example of a "living fossil" is the Coelacanth fish. Coelacanth fish fossils have been found in Pennsylvanian age rocks dating back approximately 320 million years. Long thought to be extinct, the first live coelacanth was caught off Madagascar in 1938. Such stories help to keep the study of the earth's past and its former numerous inhabitants a never-ending source of interest.

The Triassic period is considered by many paleontologists to mark the emergence of the first mammals, however little is known of their physiology as detailed evidence currently appears to be absent from the discovered fossil record. New fossil discoveries impacting the previously believed size and nature of long-extinct life on earth are made occasionally and reported in the main-stream media. Among the invertebrates or animals lacking a backbone and a bony skeleton, insects were represented in the Triassic by the first species to undergo complete metamorphosis from larva through pupa to adult. Other invertebrates of the period were the squid-like belemnites and ammonites, and the first lobsters.

The author recalls studying paleontology or the science of fossils while pursuing a geology degree in college and learning that the discovered and preserved fossilized remains of plants and animals in the rock record probably represents only a fraction of the life-forms that actually existed on earth ages ago. A rather memorable realization and sense of perspective.

Plate tectonics is a theory of global tectonics (geologic structural deformations) that has served as a key in modern geology for understanding the structure, history and dynamics of the earth's crust. The very established theory is based on the observation that the earth's solid crust is broken up into about a dozen semi-rigid plates. The boundaries of these plates are zones of tectonic activity, where earthquakes and volcanic eruptions tend to occur. One well-known boundary in North America is the earthquake-prone San Andreas Fault that emerges from under the Pacific Ocean near San Francisco and then crosses hundreds of miles inland. The San Andreas Fault has been interpreted to be part of the boundary between the North American and Pacific plates.

The plate tectonics revolution in geologic thought occurred in the 1960's and 1970's, however the roots of the theory were established by observation and deduction a century earlier. James Hall (1811–1898), a New York state geologist, observed that the sediments accumulated in mountain belts were at least ten times thicker than in continental interiors. The observation led to the geosynclinal theory that specified that continental crust grows by progressive additions that originate as ancient and folded geosynclines, hardened and consolidated into plates. The theory was well established as early as the second half of the twentieth century.

In 1908–1912, theories of continental drift were proposed by the German geologist Alfred Lothar Wegener and others, who concluded that continental plates rupture, drift apart, and eventually collide with each other. Such collisions are believed to have crumpled extensive geosynclinal sediments that resulted in the creation of mountain ranges.

The concept of seafloor spreading as a key mechanism of continental drift and mountain building was advanced in the 1920's when an echo measuring device was developed for antisubmarine defense during World War I (1914–1918) and was modified to measure ocean depths. Volcanic arcs and subduction zones relating to the dynamic geology and significant seismic tremor problems unique to Pacific coasts were recognized as early as the 1930's by American seismologists studying earthquakes. An integrated plate tectonics theory was advanced with the knowledge of seafloor spreading and subduction zones and a system of geodynamics was created.

Bibliography

American Geological Institute (1976) Dictionary of geological terms, revised edn. Anchor Press, p 221
Bates RL, Sweet WC, Utgard RO (1973) Geology: an introduction, 2nd edn. D.C. Heath and Company, Ohio State University, pp 57, 398–499
Funk and Wagnall's New Encyclopedia (1985) vol 26. Funk and Wagnall's, Inc., p 36
Funk and Wagnall's New Encyclopedia (1986) vol 21. Funk and Wagnall's, Inc., pp 67–70
Hunt CB (1967) Physiography of the United States. W.H. Freeman and Company, pp 169–172
Steele WM (1989) Petroleum exploration wells in Georgia 1979–1984, # 77. Georgia Department of Natural Resources

Chapter 5
Challenges and Logistics of Hydrocarbon Exploration in the Southeastern United States

Most of the Southeastern United States area is far from established production; however, land lease prices may be favorable. Most of the eastern seaboard and much of the lower 48 states land mass is under-explored using oil and gas exploration drilling. The sedimentary sequence is thinner than the Gulf Coast and some reservoir quality rocks were probably subjected to fresh water invasion and flushing over time. In spite of these important issues, the author is optimistic and believes that some significant hydrocarbon accumulations may still remain awaiting discovery.

The southeast United States is close to labor and capital resources, oil and natural gas pipelines and residential and commercial industrial markets. The new oil and gas field discoveries may be modest; however, the $30–$40/bbl. Oil prices experienced during early 2015 proved to be temporary. We recall that oil was nearing $100/bbl. as recent as 2014 and probable persistent future high prices per barrel for crude oil should help offset the drilling and finding costs. Natural gas prices were still low ($2.00 – $3.00/MCF) in 2015; however, they are affected by unpredictable weather and demand cycles. The demand and price for natural gas will increase during future cold weather cycles and as coal-fired electricity generating plants go off-line having been mandated by the Environmental Protection Agency (EPA) to be replaced with modernized natural gas fired plants. Natural gas demand will also increase as a greener alternative to gasoline and diesel fuel. To save space and reduce fuel tank size natural gas can be liquefied to produce liquefied natural gas (LNG) especially for large trucks and buses. The utility of propane and butane known as liquefied petroleum gas (LPG) has been known for decades for heating and cooking. Propane is especially well adapted for use as an alternative to gasoline as a motor fuel.

Competition between oil field service providers in the southeastern U.S. will be beneficial overall and decrease the time required to further evaluate the region. Limited competition currently exists and includes other geophysical contractors. One notable survey program is in progress and is planned to extend from the eastern

R. J. Brewer, *Hydrocarbon Potential in Southeastern United States*,
SpringerBriefs in Earth Sciences, https://doi.org/10.1007/978-3-030-00218-3_5

Gulf of Mexico (GOM) through the Florida peninsula in an effort to provide 2D reflection seismic images indicating the structure and stratigraphy of the Florida peninsula. Additional plans reportedly indicate an expansion into southern Georgia and an effort to combine programs with existing limited 2D seismic lines currently on the market and offered for license by data brokerage firms.

Political challenges also are known to exist in the Southeastern United States. Georgia has experienced more than one failed oil and gas exploration campaign in the past and some residents may now feel that new efforts may be a waste of resources. The current public aversion involving misinformation regarding hydraulic fracturing and its alleged adverse environmental impact is one important issue. Offshore drilling is another issue. It's certainly understandable that everyone wants ready access to plentiful and inexpensive oil and natural gas supplies to help support their lifestyles yet no one wants to see evidence of the required infrastructure in their backyards.

Bibliography

Pickens TB (2008) The first billion is the hardest. Crown Business, Random House Inc., New York

Chapter 6
Oil and Gas Production in Virginia and Tennessee

The fact that commercial oil and gas production exists in Virginia and Tennessee and even far western Maryland bodes well for possible oil and gas accumulations in the southeastern United States because the areas share similar stratigraphic and structural geology. The oil and gas fields already discovered in the western Appalachian Basin proper can serve as models or analogs to similar features being discovered elsewhere in the central and eastern parts of overlooked sedimentary basin features of the southern Appalachian Mountain system.

The oil and gas production in Virginia is located in the far western part of the state. Oil fields are located in Lee County. Gas fields are located in Scott, Washington, Wise, Dickenson, Russell and Tazewell Counties.

The nearest production to northern Georgia is in southeastern Tennessee and is an extension of the established production of northeast Tennessee located in Fentress, Morgan & Scott Counties. It was established in the mid 1800's. The lithology is Ordovician - Mississippian age fractured limestones of the Monteagle, Bangor, Lebanon, Ft. Payne, Devonian and Chattanooga Shale (source and reservoir rock.) formations. Strong unconventional shale gas potential exists involving the thick and organic rich Conasauga shale formation. New horizontal drilling and reservoir hydraulic fracturing technologies are expected to help further develop the area.

The established oil and gas production in Maryland is relatively recent and not well known and is an eastern extension part of the large Marcellus and Utica shale unconventional reservoir play that currently dominates exploration and production activity in Pennsylvania, Ohio and West Virginia and also extends into southern New York. Like other unconventional shale plays in the United States such as the Eagle Ford, Woodford and Barnett; directional drilling and hydraulic fracturing have been a boon to the Marcellus Utica play making it economically possible to get high yields from the tight shale reservoirs.

The existence of commercial quantities of hydrocarbons in Tennessee and bordering states that may be interpreted to be in the southern part or extension of the

R. J. Brewer, *Hydrocarbon Potential in Southeastern United States*, SpringerBriefs in Earth Sciences, https://doi.org/10.1007/978-3-030-00218-3_6

huge Appalachian Basin is encouraging for future exploration throughout the region. The idea of possible commercial oil and gas accumulations in Georgia and portions of the other states of the Southeastern U.S. that are on the eastern and southeastern side of the Appalachian Mountains is intriguing because the accumulations may exist in woefully underexplored and smaller hidden sedimentary basins perhaps within the Blue Ridge Thrust Belt Province and the Piedmont Province awaiting discovery by the drill bit. The smaller sedimentary basins may be hidden because of the lack of well log data and more importantly – the lack of enough high-quality reflection seismic data needed to find and map prospective subsurface structural and stratigraphic features that act as hydrocarbon traps. The features include anticlines, large sand strand lines formed from eroded rocks washing down off the Appalachian Mountains and formation pinch outs and unconformities that have formed perpendicular to regional dip. From personal experience talking to some of his colleagues, the author has concluded that it may be far too easy and convenient to shrug off and/or attempt to explain away possibilities of oil and gas existing in the southeastern United States. This challenge may be due to pre-conceived geologic thought complicated by the widespread lack of subsurface data in this vast area. We are reminded that only the drill bit finds oil and gas reservoirs.

The hydrocarbon accumulations in Virginia and especially Tennessee were also pre-dated by the observance and exploitation of oil and gas seeps and may serve as an additional model to continuing exploration efforts in neighboring Georgia. Some operators reportedly planned to drill some exploration targets in northwestern Georgia involving the Chattanooga and Conasauga shale formations. Additional prospective formations of these prospects may also involve the Monteagle, Bangor, Lebanon, and Ft. Payne formation equivalents that produce in Tennessee.

Chapter 7
Possible Undiscovered Oil and Gas Accumulations and Documented Oil and Gas Shows in the Southeastern United States

The majority of oil seeps in Georgia seem to follow a line from south of Augusta on the Savannah River, through Louisville, Sandersville, Wrightsville, lower Laurens County, Scotland and Hawkinsville, along the line separating the lower and upper coastal plain.

It is hoped that the oil seepages in Georgia are live oil and not relict oil originating and tentatively pointing the way to a substantial commercial sub-surface petroleum reservoir or reservoirs. It is hoped that the source will eventually be eventually reached by the drill bit at some, hopefully shallow as possible, unknown depth. It is fascinating that the source or sources of the oil seeps has long eluded numerous and at times very focused, scientific and very expensive attempts to find it. The oil and gas exploration giant - Chevron spent considerable amounts of time and money in Georgia gathering seismic and well log data, leasing land and drilling in the early 1980's. Traces of oil and gas have been reported in the literature discussing the region dating back from the first half of the twentieth century. Natural oil seepages occur near the town of Scotland in Telfair County, Georgia.

The oil seep near Scotland is the most famous and largest known oil seep in Georgia with the largest observed accumulation of oil. Soon after it was reported and publicized in local newspapers around 1919, a small amount of the crude oil was subjected to thorough laboratory testing and found to be similar in makeup to some of the crude oil found in California - most of which are of good refinement quality to yield gasoline and other distillates. This was encouraging news and deepens the mystery while favoring theories that the oil found is live and seeping up from the subsurface from a reservoir of accumulated hydrocarbons. The most plausible theory is that the oil is originating down dip in a reservoir that may be perhaps a sand strand line from an ancient buried beach or river bed and then slowly migrating up dip from the Coastal Plain towards the Fall Line and Piedmont. The migrating oil is probably reaching the surface via faults and fissures in the rock too small

R. J. Brewer, *Hydrocarbon Potential in Southeastern United States*, SpringerBriefs in Earth Sciences, https://doi.org/10.1007/978-3-030-00218-3_7

to be indicated by the very sparse well log data and predominately poor quality seismic data profiles acquired in the area years ago. Well logging and seismic data technology have improved considerably especially since the mid 1990's. Re-examination of well log data from wells in the general area and especially those located down dip from the seeps found in all known seep locales in Georgia along with higher quality seismic data processed with high technology that significantly improves the quality of the seismic data will help and make the data interpretation and mapping easier.

One such technology is cepstrum signal processing which exploits the periodicity or redundancy in seismic data.

Cepstrum seismic data signal processing where signal is the down going and reflected seismic energy minus the noise, is most applicable to 3D seismic data. Cepstrum seismic signal data processing is a fourier transform operation and deconvolution process that can yield multiple free seismic data. Seismic multiples are false reflection phenomena that usually occur in all seismic data and complicate the data interpretation and mapping process. The complications which may include mapping the wrong seismic event which relates to an actual subsurface rock formation can lead to erroneous conclusions unless they are minimized or eliminated from a seismic data set prior to the data interpretation and mapping process. Any modern data acquired whether 2D or 3D optimally data processed to yield the highest resolution should help image and isolate buried sand reservoirs or fractured carbonate (limestone and/or dolomite) features awaiting testing by the drill in the southeastern United States.

A summary of oil and gas shows in wells drilled in the southeastern United States follow. A test well in Georgia near Scotland was recommended in 1919 to tap reservoir rock and reached TD in crystalline basement rock that was expected @ 3000 ft. Traces of oil and gas have been reported in Cretaceous sediments at a few places, notably in a deep well drilled at Summerville, South Carolina, at one locality in Telfair County, Georgia, and more recently in a well in Lake County, Florida. Shows of oil and gas have been reported to have occurred in the following locations in Georgia: Ware County: Oil show at 2000 ft. TD 3040 ft. Jeff Davis County: Oil show in Ripley formation. TD 1975 ft. Chatham County: Gas show at 1000 ft. Oil show at 1590 ft. TD 2130 ft. Burke County: Three (3) wells drilled to 1500 ft., 1540 ft. and 1560 ft. with oil shows reported in one (1) of the wells. Montgomery County: Rumored oil show at 1542–47 ft. TD 1585 ft. Some wells drilled for hydrocarbons have established good reservoir quality rock with fresh and salt water encountered. Water being encountered in the exploration wells is discouraging; however, may indicate good quality reservoir rock that may hold and trap oil and gas accumulations elsewhere.

Shallow fresh water artesian wells exist in the region and Georgia has an abundance of high quality water wells that may indicate good hydrocarbon reservoir rock at depth. Data from water wells have been useful in the past to help map the subsurface and possibly indicated the presence of structures at depth.

Recent well test data is not published or readily available. It is suggested to check with state departments of natural resources for log data, etc. Only by continued well drilling will definite answers be obtained.

An analog or model – albeit optimistic for the accumulation of hydrocarbons in the southeastern United States and East Coastal basin applicable and relating to sand and clastic reservoir rock sourcing may be the giant East Texas oil field or pool discovered on October 3, 1930 that covers over 140,000 acres and is located in large portions of Gregg, Upshur, Rusk and Smith Counties, Texas. The field is lying over 3500 ft. below the surface under the towns of Kilgore, Overton and Gladewater. The field was discovered by a team led by wildcatter Columbus Marion (Dad) Joiner of Alabama. The classic oil field may serve as a large and very optimistic analogy of possible undiscovered fields in the onshore and offshore East Coastal Basin of the United States. The East Texas field upper Cretaceous-age (145–66 million years ago), Woodbine formation reservoir is composed primarily of relatively clean quartz sandstone. The northeast-southwest trending field extends in length for over 40 miles and varies in places from about 5–8 miles in width. It is a mildly structurally enhanced (monocline) stratigraphic petroleum trap. The trap exists at the intersection of two unconformities (erosional and once exposed at the surface) planes along and against the eastern and up-dip edge pinch-out of the Woodbine reservoir sand. The very-productive Woodbine reservoir quartz sandstone is overlain by the Eagle Ford shale, Austin chalk and Annana chalk formations which also act as a seal and barrier to oil migration and escape. The Eagle Ford shale is believed to be the source rock for the oil accumulation. The Woodbine reservoir sand is underlain by the Washita and Fredericksburg formations. The average thickness of the Woodbine reservoir sand in the East Texas pool is about 35 ft. Dominate geologic thought is that the prolific and productive quartz-rich Woodbine sands were most probably sourced by streams flowing south toward the ocean after cutting through and eroding the Ouachita Mountain range of Oklahoma approximately 200 miles directly to the north. Over 5.42 billion barrels of oil have been produced from the field. The field may have originally contained in excess of 7 billion barrels of oil.

Another postulated source for the Woodbine quartz sands present in the East Texas oilfield is from the huge southern Appalachian Mountain complex well over 500 miles to the northeast. The Appalachian Mountain sourcing theory is not nearly as realistic because of the much greater and tortuous quartz sand transport distance via streams involved. Regional dip tectonics and sedimentation history of the southeastern United States during Cretaceous geologic time does not favor the Appalachian Mountain sand sourcing theory for the East Texas oil field.

Quartz sand being transported south and east from the ancestral (and considerably higher altitude), Appalachian Mountain range during the Cretaceous Period of 145–66 million years ago would most probably be transported by eroding streams flowing south and east on their way to the ocean. It is hoped that large quartz-rich and permeable sandstone hydrocarbon reservoirs remain hidden a few thousand feet under the east coastal plain of the eastern United States and perhaps also buried under the offshore shelf edges awaiting discovery by the drill. The traps are probably stratigraphic in nature and not readily imaged and able to be located by seismic data. Clastic reservoirs may also be located under over thrusted strata nearer to Appalachian mountain foothills and even perhaps further to the northeast. Sparse 2D reconnaissance reflection seismic profiles completed during the early to middle 1980's including the COCORP data set indicate the possible presence of the over thrust features. Much more data is needed to help locate exploration drill sites to substantiate the theories.

Bibliography

Postley OC (1938) Oil and gas possibilities in Atlantic Coastal plain from New Jersey to Florida. AAPG Bull 22(7):799–815
Thompson S (2008) Black gold! Texas tea! Oil in Georgia. Editorial, The Courier Herald, Dublin

Chapter 8
Historic COCORP 2D Seismic Survey Program in the Southeastern United States

The COCORP 2D surface seismic data survey program was an inspiration to the author to produce this work. The COCORP data located in the state of Georgia is especially intriguing and thought-provoking as it appears to have imaged some large sub-surface structures that may include anticlines with essential four-way hydrocarbon trapping attributes that remain untested by the drill bit.

The COCORP project was funded primarily by the National Science Foundation (NSF). Truck mounted vibrator energy sources were used. The most comprehensive COCORP program is in Georgia. The key project objectives reportedly were to image the Mohorovicic discontinuity (Moho) and the North American tectonic plate boundary or suture. The Mohorovicic discontinuity is believed to be a seismic discontinuity situated about 35 km below the continents and about 10 km below the oceans which separates the earth's crust and mantle. Additional objectives were to image basement thrust faults and indicate evidence for Triassic rift basins. A 15 second record listen time, 24–48 common depth point (CDP) fold and off end and split spread recording geophone spread geometries were used. It was a very successful project; however, needs data reprocessing. Higher fold 2D coverage is needed. The original traverse needs to be re-acquired and expanded to tie key wells. Some structures imaged on the original survey still remain to be tested in the northern part of Georgia. Some anomalies are deep; however, appear to be within the reach of modern drilling technologies. Some COCORP lines suggest sediments may exist below basement crystalline rocks. Strong evidence of over thrusted structures is indicated. The concept and idea is expandable south to Florida and northeast to South Carolina, North Carolina and Virginia. A limited number of COCORP lines exist in limited parts of Florida, Tennessee and South Carolina.

A northwest to southeast geologic cross section traversing about 75 miles located in Tennessee, North Carolina and Virginia illustrates a series of sub-horizontally westward – overriding fault sheets, also termed allochthons or nappes. Older strata above an over-thrust are repeatedly superimposed on younger strata. The interpretation of COCORP reflection data farther east indicates that one or more major thrusts

R. J. Brewer, *Hydrocarbon Potential in Southeastern United States*,
SpringerBriefs in Earth Sciences, https://doi.org/10.1007/978-3-030-00218-3_8

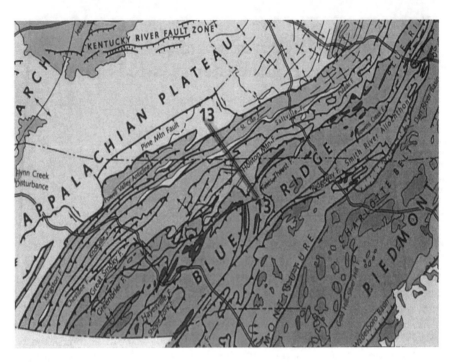

Fig. 8.1 Traverse of NW-SE cross section in North Carolina, Virginia and Tennessee. Line 13

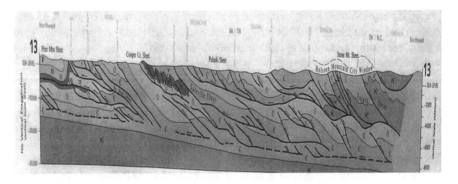

Fig. 8.2 NW-SE cross section in North Carolina, Virginia and Tennessee
Potential hydrocarbon accumulations may be present in over-thrusted and folded strata. A series of 2D reflection seismic profiles will help to confirm extent of over-thrusted and folded rocks

dip gently eastward below the Piedmont area throughout much of its width. These data also suggest that younger, un-deformed and possibly oil and gas bearing early Paleozoic strata may lie within reach of the drill below the western portion of the highly deformed crystalline rocks of the Blue Ridge and western Piedmont areas (Figs. 8.1 and 8.2).

At some point an as long as possible portion of the 75 mile long cross section traverse will need to be substantiated with a seismic data profile over the same general NW-SE traverse that will effectively prove the validity of the older strata above an over-thrust model. Ideally a series of at least six seismic data profile traverses in the same general vicinity will greatly aid to the knowledge of the subsurface and will probably locate a number of possible drilling targets.

Bibliography

American Geological Institute (1976) Dictionary of geological terms, Revised edn. Anchor Press, p 284

Cook FA, Brown Larry D, Kaufman S, Oliver JE (1983) The Cocorp seismic reflection traverse across the southern Appalachians, AAPG Studies in Geology No. 14. The American Association of Petroleum Geologists, Tulsa, p 61

Chapter 9
Georgia as a Central Oil and Gas Exploration Project Area

In 1971, Corpus Christi, Texas geologist Richard H. Sams visited and examined the sites of eight hydrocarbon seeps in the state of Georgia and published his findings in a very informative and useful report that reviewed the same seeps documented 50 years before in the early part of the twentieth century. Sams did not visit the site of a ninth seep located near the town of Cedar Springs in Early County, Georgia in the far western part of the state. He clearly states in his report that the Cedar Springs seep was not visited because it was too far away from the other seeps that are located much farther east and closer to each other in the central part of Georgia south of the Fall Line. This is intriguing because the author was not aware of the existence of the Early County seep before purchasing a copy on eBay and then later reading the Sams report in April of 2015.

The author also was not aware of the existence of the Unadilla seep of Dooley County, Georgia, the Hartford seep of Pulaski County, Georgia and the Augusta seep of Richmond County, Georgia. After plotting all known seeps in Georgia on a map greatly aided by the map details contained in the Sams report, it became surprisingly evident that the albeit remote western Georgia Cedar Springs seep, is basically on strike or that is to say right angle to dip, with all the other known seeps in the state. Regional dip in the coastal plain of Georgia is southeast toward the Atlantic Ocean. The strike of the seeps appears to be of course at a right angle to regional dip and trends to the northeast. Sams' admirable reinvestigation efforts were complicated by 50 years of erosion, reforestation and relocation of highways and landmarks.

Sams' findings unfortunately concluded that most of the seeps reported to be active around the early 1920's and way before had dried up including the most famous and storied seep located near the town of Scotland in Telfair County. The inactive locations have been interpreted as relict hydrocarbon seeps. Fortunately, the Wrightsville seep located in Johnson County, Georgia and the Louisville seep located in Jefferson County, Georgia appear to have withstood the test of time at

49
R. J. Brewer, *Hydrocarbon Potential in Southeastern United States*,
SpringerBriefs in Earth Sciences, https://doi.org/10.1007/978-3-030-00218-3_9

least up until 1971 and still indicated evidence of hydrocarbon seepage per the Sam report.

Sams provides a particularly interesting anecdotal account involving the Sandersville seep located in Washington County, Georgia. He reported that he was told by a local resident who remembers playing very near the seep 20 years earlier circa 1950 and describing the oil encountered as being thick and bluish in color. The seep was reported to be on very high ground in a road cut which passed down the hill and near the local resident's house. Sams did more research and found that newspapers on file in Sandersville indicated that the seep in question was discovered by a work crew in 1920 surveying and cutting a new road through the area. The newspaper account reported that the work crew noticed some gas associated with the oil seep at that time. In January of 1921, the Middle Georgia Oil & Gas Co. apparently tested the seep with a shallow well – the Lillian B. No. 1 drilled to 605 ft. The well was a dry hole, staked about 10 miles southeast of the Fall Line with no reported shows of oil and/or natural gas. The well was TD'd in reportedly a "basement complex".

The author is of the opinion based a very limited information, that the Lillian B No. 1 location bore hole may not have been drilled deep enough to adequately prove up the noteworthy oil seepage evidence reportedly discovered at the surface by eye-witnesses dating back to the first half of the twentieth century and probably much earlier. The "basement complex" term may be a reference to crystalline igneous rocks below the sedimentary rocks. Some theories indicate that additional prospective sedimentary rock sequences may exist under igneous rocks near this part of Georgia that may have been thrust over older sedimentary rocks as evidenced by reflection seismic profile surveys conducted in the 1980's.

As of 1971, the reported seep area near Sandersville no longer shows signs of oil seepage and the road cut that was responsible for the discovery of the seepage has now reportedly been backfilled covering all evidence or prior oil seepage! Even more surprising is that Sams goes on to report that the original road itself has been rerouted so that earth has been pushed over the original seep area. The newer excavation for the road is located a very short distance downhill, but it contained no indications of hydrocarbon seepage. One might wonder what may be re-discovered if the original seepage area is excavated.

A conceptual and idealized geological cross section may serve to shed some light on where the small oil seeps in Georgia may be sourced from. Larger oil accumulations may be down dip or up dip from the small oil seeps. Down dip subsurface oil accumulations may nearer to the coastline or offshore. Up dip accumulations may be trapped in buried sediments up under the Appalachian Mountain foothills or mountain fronts. Further investigation is needed to determine if additional oil seeps are occurring on the ocean floor offshore Georgia or up in the Piedmont and Appalachian Mountain front of the northwestern part of the state. Fig. 9.1.

A re-investigation of all known oil and gas seeps in Georgia and throughout the southeastern United States is clearly warranted to get more details and see if some relict seeps have become active again. There is also the possibility that new seep locations may be discovered and previously located seeps will become re-activated.

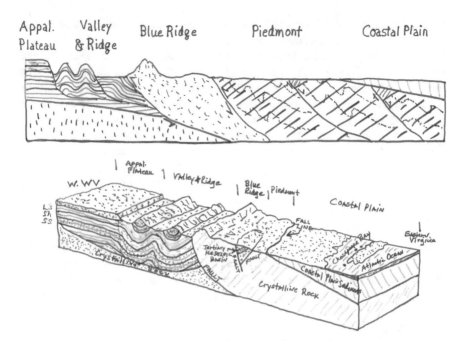

Fig. 9.1 Block diagram and generalized cross section from Appalachian Plateau to Coastal Plain

Ground water springs have been known to dry up and then re-activate and flow again during seasonal changes. In some cases, if a person goes to work and intervenes with in some instances only a pick and shovel, the spring may start to flow again anew. Ranchers and farmers in hilly and rocky terrain have known this trick and have used it in arid climate locales like New Mexico.

Notable oil companies that have drilled in Georgia and Florida chasing seeps include Chevron that drilled one of the deepest tests in Jeff Davis County, Georgia to a TD of 11,470 feet. in 1981. Exxon (Humble Oil), Hunt Oil Sonat (El Paso) have also drilled in Georgia. BreitBurn Energy Company developed some of the Sunniland (Cretaceous) oil field in Lee, Hendry, Collier and Dade Counties, Florida. Other companies to drill in Georgia were Pan American (Amoco), (BP), Houston Oil and Minerals (Seagull Energy Corporation), Kerr-McGee (Anadarko), South Eastern Exploration & Production Corporation, Sun Oil Company (Oryx), Kerr-McGee (Anadarko), Pennzoil (Texaco, then Chevron), and Mobil (Exxon). Obtaining critical data from their past efforts will be very challenging due to the proprietary nature of these data that were obtained at considerable time and expense. The never-ending cycle of corporate buy-outs, mergers and acquisitions will present additional obstacles. Some of these old data were reportedly destroyed and/or misfiled and therefore perhaps lost forever.

Fast forward to October 2015. Georgia newspapers including the Atlanta Journal-Constitution have reported that oil and gas exploration operators and companies are leasing land and have drilled a few wells in northern Georgia and have had encouraging results in the Conasauga shale formation.

The Conasauga shale formation is reportedly one of the thickest shale sequences in the world and may extend to as much as 15,000 feet in total thickness in some locations. The shale formation is Cambrian in age and is part of the Appalachian thrust and fault region of Alabama, Georgia and Tennessee. The Cambrian period in geologic time is quite old and is estimated to have occurred 541 to 486 million years ago. The Conasauga shale is present as far south and west as Alabama and extends well up to the northeast into far eastern Tennessee and northwest Georgia. The shale is productive in Alabama and has reportedly has been encountered in exploration wells as shallow as 4660 ft. and as deep as 12,286 ft. The shale is overlain by the Rome formation which is in turn overlain by the Knox formation. The Knox formation is a dolomite and limestone sequence and produces commercial quantities of hydrocarbons in Tennessee. The Rome formation is a shale, siltstone and sandstone sequence and produces in Kentucky and Tennessee. Natural gas accumulations from the Conasauga shale have been notably demonstrated in Alabama as early as 1984 and as late as 2009. The total recoverable amount of natural gas contained in the entire extent of the shale is estimated to be very high and into the billions of cubic feet.

Natural gas has reportedly been found in potentially commercial quantities and will eventually be encouraged and supported by higher natural gas prices. A ready market for the gas exists in the region. In many cases, the Conasauga shale natural gas reservoir will probably need to be hydraulically fractured in order to produce the high enough flowing natural gas yields required much like the Bakken shale gas bonanza of North Dakota.

The Bakken shale formation natural gas field in North Dakota ranks as one of the world's largest fields. The Bakken is found primarily in North Dakota, however; also extends well northward into Canada and is located in the very large Williston Basin petroleum province. The success of the Bakken play is due in large part to significant technology achievements including directional drilling and proven reservoir stimulation techniques such as hydraulic fracturing. The Bakken shale play has been a huge boon to the local economy in and around North Dakota by creating many high paying jobs reportedly ranging from $50,000 to $100,000 per year and higher depending on the type of job. Unfortunately, low natural gas prices starting around 2013 have been a significant damper on the Bakken shale play as well as all natural gas well drilling and field development nationwide. Hydraulic fracturing works very well in brittle tight formations that do not flow hydrocarbons easily such as the Bakken shale and the Barnett shale of Texas. Directional drilling and hydraulic fracturing were perfected by a team led by the renowned oilman, real estate developer and philanthropist George P. Mitchell (1919–2013), starting in the late 1970's, and through the 1980's and 1990's. Mitchell is credited with pioneering and economically proving up the technologies with great success in extracting hydrocarbons in specifically the Barnett shale, Austin Chalk and Cotton Valley formations of east Texas and the Fort Worth basin petroleum province.

An adequate pipeline infrastructural currently does not appear to exist in the areas of the Conasauga shale of the southeastern United States and will have to be constructed. Parts of South Carolina including the affluent Hilton Head Island area must rely on propane gas being transported in by truck. Residents are forced to pay premium prices for home and business heating needs.

Other parts of the United States including the remote Upper Peninsula of Michigan are far away from municipal natural gas pipeline networks and so must rely on expensive LPG (liquified petroleum gas) propane and butane for heating needs. More natural gas pipelines need to be constructed in remote areas when the population increases to justify the costs involved.

Ironically, Great Lakes Transmission Company (GLGT) operates a high pressure natural gas pipeline going across the Upper Peninsula of Michigan through Schoolcraft County, Michigan and near residential areas that must use LPG for heating.

Heating using electricity is usually cost prohibitive. Environmental concerns over hydraulic fracturing operations potentially impacting fresh water aquifers is a significant issue in the southeastern United States and has also become a nationwide concern. The new high-paying jobs resulting from natural gas production, however; will be a very welcome and long overdue boon to the Georgia economy and the entire region.

In 1919 the Geological Survey of Georgia with a team led by State Geologist S.W. McCallie conducted an innovative way to search for subsurface anticlines that may act as a trap for hydrocarbons in the vicinity of the well-known oil seep near Scotland, Georgia by mapping the Tertiary Oligocene Chattahootchie limestone that was penetrated by a series of water wells on a northwest to southeast traverse from Helena through Scotland to Lumber City, Georgia. Four anticlines were indicated from the work that may have been later tested by unsuccessful drilling. It would be a good idea to conduct this low-cost exploration method in other areas to indicate additional anticline structures. It is not known if other water well anticline mapping work like this was done anywhere else.

Formation top maps can be made from the formation tops taken from electric wire line well logs of the Lower and Upper Cretaceous, Tuscaloosa and Eutaw formations. The compiled formation top information is available from publications produced by the Georgia Department of Natural Resources Georgia Geological Survey.

Most exploration wells in Georgia were drilled in the early to middle twentieth century. Very little well drilling occurred since the 1980's. More wells need to be drilled in Georgia and the entire southeastern United States because the subsurface control needed to make more accurate structure maps from formation tops still does not exist to readily reveal where to drill new oil and gas exploration wells.

Very little well test data exists from key potential producing formations such as the Tuscaloosa and Eutaw formations from the wells drilled in the twentieth century, yet natural gas shows have been reported in addition to hot salt water flowage from the 3007 ft. J.R. Sealy Fee #1 location in Dade County, northwestern Georgia drilled in 1953 that indicates possible geothermal potential and good quality reservoir rocks.

The average bottom hole temperature (BHT) of all wells studied needs to be re-examined in the area to aid the planning of future well drilling and formation evaluation logging work.

Sample logs from all available wells need to be re-examined for clues to possible by-passed oil and/or gas bearing zones. The search for high quality reservoir sandstones like quartz sands will be continued to try and find large sandstone reservoirs like the Woodbine quartz sandstone formation at the large East Texas field.

The author believes that eventually a commercial find of natural gas and/or oil will be discovered in Georgia.

Four (4) publications were published and may be still available from the Georgia Department of Natural Resources and include Oil Tests in Georgia by Vernon J. Hurst, Information Circular 19, June 1960, Oil Tests in Georgia – A Compilation, A.S. Furcron, Director, Information Circular 19, third Edition, June 1965, Petroleum Exploration Wells in Georgia, compiled by David E. Swanson and Andrea Gernazian, Information Circular 51, 1979 and Petroleum Exploration Wells in Georgia 1979–1984, Information Circular 77, Compiled by William M. Steele, 1986.

The first three publications give formation tops from sample logs including the base of the Cretaceous, top of the Tuscaloosa, top of the Eutaw and the top of the Cretaceous as well as information about what type of electric logs were run in each well. The author was able to produce contoured structure maps of the base of the Cretaceous, top of the Tuscaloosa, top of the Eutaw and top of the Cretaceous. The fourth publication indicates the type of electric logs run in each well but does not give formation top or base depth information from sample log or electric logs and was of very limited aid to the author's mapping work. Unfortunately, indications are that some of the operators of the wells listed in Petroleum Exploration Wells in Georgia 1979–1984 may have chosen to not report formation tops and treated their locations as tight holes. It is unknown as to why formation top information is not listed because tight hole information regarding formation tops expires after a while and should have been included.

The structure maps of the base of the Cretaceous, top of the Tuscaloosa, top of the Eutaw and the top of the Cretaceous all indicate an interpreted network of wrench faults at each interval that interrupts regional dip and may help explain why operators located wildcat wells where they did. Opportunity for additional drilling may be indicated by trying to get structural advantage to the dry hole locations. Not surprisingly, the top of the Cretaceous structure map shows the least amount of regional dip interruption while the base of Cretaceous shows the most amount of regional dip interruption as it is closest to the structural forces that affected the igneous crystalline basement rocks.

It will be interesting to eventually compare the sample log tops used by the author to construct his structure maps to the set of log tops obtained from electric well logs. The structure maps made from electric log tops would be expected to be more accurate and may not indicate as much faulting and/or lower magnitude of fault throw or displacement between wells.

Prior to 1979 the deepest wells drilled in Georgia went to about the 4000–6000 ft. depth range where crystalline basement rocks were encountered and it was felt unnecessary to drill deeper because of lack of additional prospective sedimentary rock section. After 1979 a few wells were drilled much deeper. A very notable example is the Chevron A.P. Snipes et al. No. 1 location in Jeff Davis County, Georgia that was completed on June 29, 1981 to a depth of 11,481 ft. The Snipes location may still have the distinction of being the deepest well drilled in Georgia to date.

In 1923, the Geological Survey of Georgia published <u>Petroleum and Natural Gas Possibilities in Georgia</u> bulletin No. 40 by T.M. Prettyman and H.S. Cave, 164 pages. A map was included in the book between pages 134 and 135 presenting a geologic map of the coastal plain of Georgia including a structure map of the base of the Tertiary Miocene Alum Bluff formation.

Using well tops of the base of the Alum Bluff formation furnished by the map the author reproduced a structure map of the base of the Alum Bluff formation in an effort to determine whether or not the Alum Bluff formation structure map would indicate structural features deeper in the subsurface including at the top of the Cretaceous, top of the Eutaw, top of the Tuscaloosa and base of the Cretaceous formations. Two structural nose features are indicated from the base of the Alum Bluff formation structure map. One structural nose or high is centered in Coffee county and another is centered in Candler county. The structural noses will be correlated with the deeper structure maps prepared by the author and investigated further.

Structurally high areas found from mapping the top of the Cretaceous formation have been indicated and may be located in Charlton, Effingham and Marion Counties, Georgia.

Apparent structural highs in the top of the Eutaw formation may be located in Dodge and Liberty Counties, Georgia.

The top of the Tuscaloosa formation may have formed structurally positive areas in Burke and Macon Counties, Georgia.

Finally, the base of the Cretaceous formation seems to have one large structurally high area centered in Houston County, Georgia.

The possibility exists that some features remained untested by drilling and it may be possible to get structural advantage to dry holes downdip. Low areas will also be investigated as possible accumulation sites for clastic and sandstone reservoir accumulations.

Seismic data will be required to help delineate and map prospective features in more detail.

Bibliography

McCallie SW, Hull JPD, Teas LP (1919) A preliminary report on the oil prospect near Scotland Telfair County, Georgia. Geological Survey of Georgia, Atlanta, p 23

Sams RH 1971,Oil seeps in Georgia, A reinvestigation. The Geological Survey of Georgia, Department of Mines, Mining and Geology, Atlanta, p 18
Swanson DE, Gernazian A (1979) Petroleum exploration wells in Georgia. Georgia Department of Natural Resources, Georgia Geological Survey, Information Circular 51, Atlanta, pp 61–67

Chapter 10
Gulf of Mexico Basin and Atlantic Coastal (East Coastal) Basins

It is not readily apparent, unless one is used to looking at regional geology maps of the United States that show the many sedimentary basins throughout the country, that the well-known Gulf of Mexico Basin converges somewhere in Georgia and Florida. with the less known and far less publicized Atlantic (East) Coastal Basin of the southeast United States in the states of Florida and Georgia (Fig. 1.1). It is also interesting to note that the Atlantic Coastal Basin, like the Gulf of Mexico Basin is very extensive and a large part extends offshore.

Exactly where as far as being able to draw a generalized demarcation line or zone between the two basins is concerned, is not yet known. In the author's opinion, because of an obvious lack of well log data and primarily from a dearth of much-needed seismic data profiles to aid in an attempt to tie the sparse well log data together, defining an accurate transition zone on a map between the East Coastal and Gulf of Mexico basins is not yet possible. Most of the wells in Georgia and Florida are not very deep (less than 10,000 ft.), therefore the limited wireline formation evaluation logs ran in these relatively shallow correlation wells hinders rock correlation and mapping efforts. By contrast, the Gulf Coast basin has many more wells. Many of the wells have reached total drill depths far in excess of 10,000 ft. The corresponding wireline logs for the wells are more numerous and detailed as a rule. The seismic data used to tie (correlate) the wells together significantly aids rock identification and mapping efforts to find and further develop the oil and gas fields found as a result (Fig. 10.1).

Virtually all of the reflection profile surface seismic data acquired from the late 1930's to the early part of the twenty first century is two dimensional (2D). 2D seismic data produces a narrow (few hundred feet) acoustic profile of the subsurface extending down thousands of feet into the rock formations basically directly underneath the recording geophones placed in a line at the surface. The line at the surface is surveyed ahead of time and can extend for many miles. Line length is theoretically unlimited. Realistically, total line length becomes quickly limited by topographic and cultural limitations, land permit issues and primarily costs. The depth

© The Author(s), under exclusive licence to Springer Nature Switzerland AG 2019   57
R. J. Brewer, *Hydrocarbon Potential in Southeastern United States*,
SpringerBriefs in Earth Sciences, https://doi.org/10.1007/978-3-030-00218-3_10

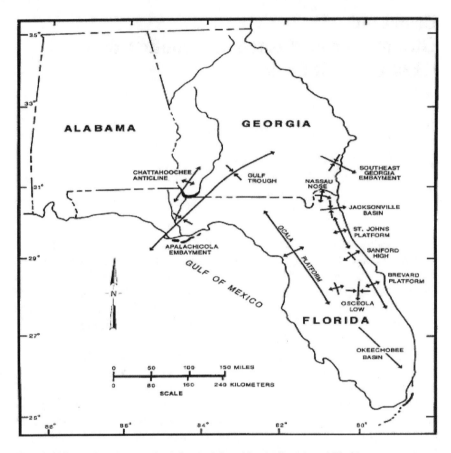

Fig. 10.1 Subsurface features that influenced deposition in Georgia and Florida

that usable and interpretable (able to be mapped), data can be acquired is limited by the signal to noise ratio or acoustic transmissivity of the total rock volume to be investigated and the strength of the energy source used. The energy sources typically employed is a very limited high explosive charge in a pre-drilled shot hole ranging from 25 ft. or less to several hundred feet in depth. The other energy source choice that is becoming much more common and of less environmental impact is the vibrator unit that is typically mounted on a truck. Several vibrator units are used at once to produce more seismic energy as their output through a base plate contacting the ground is directed down into the subsurface as the total energy output from all units are summed together.

Three dimensional (3D) seismic data started to be utilized by seismic data acquisition field crews in a commercial production mode in the 1970's and was initially severely limited by data acquisition hardware and software limitations and computing power. 3D data is now preferred over the much more limited 2D data option; however, 3D data is more costly to acquire and is generally most cost-effective when it is acquired over producing oil and gas fields. The 3D data volume then becomes a significant aid in developing the fields to maximize the estimated ultimate recovery (EUR) of the fields. EUR for an oil field is measured in barrels and EUR for a natural gas field is measured in cubic feet.

The seismic data required on a grand scale in the East Coastal basin to help unlock its geological secrets will probably have to be 2D data because of the long lengths required and the costs. 3D data will be utilized later after producing oil and gas fields are discovered and developed. The use of 3D data is usually preferable anywhere as long as it is cost-effective. It is expected that as 3D data acquisition and processing workflows continue to rapidly improve, so will its cost-effectiveness.

2D and primarily 3D data acquisition and processing is also conducted offshore in a marine environment with great success utilizing a set of geophone streamers towed behind a ship. The first 2D marine seismic data programs were commenced in the 1940's perhaps even during the second world war (WWII) (1939–1945). Even though it was lower technology by today's standards, marine seismic data acquisition was successfully utilized during the first and second world wars to detect the approximate location to aid in the destruction of enemy submarines.

3D seismic data acquisition and processing was started commercially offshore in earnest in the late 1970's. The primary energy source used in marine seismic data acquisition is the air gun impulsive energy source powered by compressed air generated by an on-board air compressor network. Because in part of the impulsive and abrupt bubble and water pressure waves produced by the air gun energy sources used in marine seismic data acquisition, other less intrusive energy sources such as the marine vibrator were developed in the 1970's and have been used in primarily shallow water exploration programs with encouraging success. An advantage of a vibrator energy source is that the seismic data acquisition technicians and operators can control the output seismic frequency measured in hertz (Hz) from the vibrator to more closely match the frequency range spectrum returnable by the earth's subsurface when the reflected signal reaches the surface to amplified and then recorded. A land based vibrator energy source operates in a very similar way.

A main driver that started to come to the forefront in the late 1990's to move from air guns to perhaps a marine vibrator system is to limit the potential harm to the hearing and other activities of marine mammals including whales, porpoises and seals. The science and best practices is commonly known as marine mammal mitigation and is employed by marine seismic data acquisition crews on a global basis to help protect and preserve marine eco-systems.

The Atlantic Coastal Basin appears to be much more narrow and elongate than the Gulf of Mexico basin and covers an area of approximately 1110 thousand sq.km. By comparison, the nearby and well-known hydrocarbon - rich northern Gulf of Mexico basin covers a not much larger surface area of approximately 1300 thousand sq. km. not including the southern Gulf of Mexico area of offshore Mexico. The major sedimentary rocks encountered in the Atlantic Coastal Basin have an estimated total thickness of 7620 meters and range in age from Jurassic to Tertiary. By comparison, the Gulf of Mexico basin which is a very large part of the Gulf Coast geosyncline has a significantly larger estimated sedimentary rock thickness of 13,000 to 15,000 meters and also range in age from Jurassic to Tertiary.

The petroleum reservoir lithologies expected in the Atlantic Coastal basin are limestones and dolomites of primarily Cretaceous age. As the major sedimentary rocks encountered in the Gulf of Mexico basin have an estimated thickness of 13,000–15,000 meters and range from Tertiary to Jurassic in age with primarily sandstone petroleum reservoir lithologies, this is one reason why the Gulf of Mexico basin is the most prolific hydrocarbon producing province of North America. The knowledge of the lithologies or details regarding the types of rock were established over many decades of oil and gas exploration drilling dating back to the first part of the twentieth century.

A modest amount of primarily 2D marine seismic data has been acquired in the offshore area of the East Coastal basin involving the United States Geologic Survey (USGS) and private oil companies.

In the United States, based on oil and gas production, the region around the Gulf of Mexico is ranked first in size, the second largest region is west Texas and eastern New Mexico – collectively known as the Permian Basin and the third largest is the Mid-Continent region centered in the state of Oklahoma and extending down into northern Texas, up into southern Kansas, east over into western Arkansas and southwestern Missouri and then west into the Texas panhandle and western Kansas.

On January 28, 2015, it was reported in the New York Times that the Obama administration announced that a large portion of the eastern coast of the United States would be considered to be opened up for the leasing of offshore federal tracts ahead of additional oil and gas exploration efforts.

The Obama administration at long last during the week of March 13, 2016 indicated that it would seriously consider approving offshore oil and gas leasing of the east coast of the United States, however; on March 15, 2016 reversed itself indicating that the Environmental Protection Agency (EPA) would not allow leasing off the east coast. The Obama administration reportedly reversed itself yet again later in the summer of 2016 caving in to pressure from environment groups and placed the entire huge tract of federal waters of the east coast of the United States off limits to oil and gas exploration.

With the election of President Donald Trump in November of 2016 – it remains to be seen how much of an impact the Trump administration will have on the United States oil and gas industry although early indications are favorable. Trump has indicated that onerous regulations that had in the past discouraged domestic oil and gas exploration will be reduced and that new regulation creation will be significantly reduced. Trump has also made it clear that American labor, products and services will be favored and incentivized instead of outsourcing and bringing in resources from overseas. A similar strategy worked in the past under President Reagan during the 1980's and should boost economic growth that had been unfortunately allowed to stagnate under the Obama administration.

The oil and gas leasing approval announcement from the Obama administration was expected to come a few years earlier than 2016, however was undoubtedly delayed by the Macondo well blowout of the Deepwater Horizon offshore drilling platform disaster of April 20, 2010 involving British Petroleum (BP). The accident tragically resulting in fatalities, spawned additional offshore drilling regulations designed to prevent similar tragedies in the future. The ruptured Deepwater Horizon well head on the sea bottom of Mississippi Canyon Block 252 is in about 5000 ft. of water. The very deep water depth was a very formidable obstacle to finally and successfully capping the wild well. Success was greatly aided by the drilling of a nearby formation pressure relief well, on September 19, 2010 - about 5 months after the accident.

The Macondo accident was a public relations nightmare and effectively set the domestic and global oil and gas industry back from a public image perspective for years to come. The Deepwater Horizon incident is considered to be one of the worst oil spills in history and reportedly released 4.9 million barrels of crude oil into the Gulf of Mexico. For historical perspective and comparison, the Ixtoc oil spill disaster occurred much farther away in the Bay of Campeche of the southern Gulf of Mexico off the coast of Mexico on June 3, 1979. It was also a well blowout and involved Pemex – the national oil company of Mexico. Until the Deepwater Horizon accident, The Ixtoc incident was also considered among the worst oil spills in history releasing an estimated three million barrels of crude oil into the southern Gulf of Mexico. In stark contrast to the Deepwater Horizon spill, the Ixtoc Sedco 135-F drilling platform was in only 165 ft. of water yet was not successfully capped until March 23, 1980, nearly 9 months after the accident. The author remembers the Ixtoc accident newspaper and television coverage portraying a situation of the vast waste of a precious natural resource and an environmental disaster of seemingly epic proportions at the time. Fortunately, no loss of life was reported from the Ixtoc accident.

To place things in more historical perspective, oil has been and currently is being released into seawater marine environments. The on-going natural process has been observed for centuries. One notable place is offshore southern California. It is unknown exactly how much crude oil has been released into marine environments in this manner over the vast expanse of geologic time. Oil tanker spills whether accidental or as a result of warfare have also had significant impacts. Submarine and surface naval warfare and airplane attacks on vessels during the first and second world wars in all ocean battle theaters resulted in the spillage of unknown amounts of crude oil and refined product into the sea.

The areas of offshore eastern United States have long been considered prospective for the existence of undiscovered oil and gas accumulations. This is evidenced by the maps produced from interpreted marine seismic surveys that were conducted by the domestic oil and gas industry as well as the USGS branch of the federal government and shared with the entire industry as part of the public domain funded by U.S. taxpayer dollars in the 1970's and 1980's. Sub-commercial reservoirs of oil and natural gas were discovered in the 1970's and 1980's off the coast of southeastern Florida and off the coast of New Jersey. Well-known commercial discoveries were also made by the Canadian government-subsidized Canadian oil and gas industry off the coast of Nova Scotia during the latter part of the twentieth century. The discoveries are solid evidence that more commercial oil and natural gas discoveries await the drill offshore from the coast of eastern Florida all the way up north and well into Canadian waters. Onshore areas of the southeast United States will also continue to be prospective and will be buoyed by oil and natural gas discoveries offshore and perhaps vice versa.

The areas include offshore tracts off the coastlines of the states of Georgia, South Carolina, North Carolina and Virginia. This is very encouraging news for the prospect of continued United States oil and natural gas energy independence. The Obama administration indicated that the plan was part of a "balanced" energy policy. During the same announcement of January 28, 2015, the encouraging news was tempered by the clear indication that long-known prospective areas in Alaska would be placed off limits to future oil and gas exploration efforts. The leasing efforts will bring in a lot of funds that are sorely needed by the heavily debt-laden United States government. Funds from the leasing efforts will also benefit the state governments involved in the areas.

Bibliography

Guoyu L (2011) World atlas of oil and gas basins. Wiley-Blackwell, Oxford, p 361

Chapter 11
Nearest Oil and Gas Production to Georgia, South Carolina and North Carolina

Regarding North Carolina oil and gas occurrence; although natural gas and oil are known to occur in North Carolina, they are not currently produced commercially in the state. Additional work is needed to evaluate their economic value. Directional drilling and hydraulic fracturing may become more common in North Carolina as the economic benefits become more evident. Between 1925 and 1998, 128 petroleum exploration bore holes have been drilled in North Carolina. None were commercial. Recent exploration has been focused in Lee and Chatham counties. As of this writing two of these exploration wells remain under permit in Lee County but are not in production. The Deep River Coal Field is located in Chatham, Lee and Moore Counties, North Carolina. Oil bearing shales are exposed in the Deep River Valley. Oil and gas shows have been reported in wells drilled in Lee County, North Carolina such as the Butler # 3 well location.

Several areas in South Carolina are considered to have potential to produce oil and gas. Interestingly, South Carolina appears to have more natural gas pipelines than North Carolina in spite of the fact that North Carolina has more potential to produce hydrocarbons at present. It would seem from the intricacy of the natural gas pipeline in South Carolina compared to North Carolina that South Carolina has a much more mature market for natural gas consumption than North Carolina. The main area is the outer Coastal Plain. It contains a relatively thick pile of sedimentary rocks including some excellent reservoir quality rocks, but hydrocarbon source rocks may be scarce. Shale sequences will need to be studied further as both a source and a reservoir rock in North and South Carolina and the entire southeastern United States as has been done in successful petroleum exploration projects such as the Barnett shale in east Texas and the Marcellus shale and Utica shale in western Pennsylvania and eastern Ohio. Seismic surveys in the Blue Ridge region suggest sedimentary rocks may be present deep beneath the crystalline rocks and may be similar to oil and gas-bearing strata in the Valley and Ridge Provinces of Virginia and West Virginia. Detailed studies have not been conducted to verify the seismic surveys. Oil and gas exploration in South Carolina occurred in two distinct periods.

R. J. Brewer, *Hydrocarbon Potential in Southeastern United States*,
SpringerBriefs in Earth Sciences, https://doi.org/10.1007/978-3-030-00218-3_11

The first period occurred between 1920 and 1957 11 wildcat wells were drilled in the Coastal Plain of South Carolina and all the wells were dry. The deepest well was the first well drilled in 1920 or 1921 near Summerville to depth of 2570 ft. The Summerville well reportedly found traces of oil and gas. The second and current period of activity began in 1980. These activities included leasing, soil geochemical analyses, seismic surveys, and drilling, and major oil companies and small independents have conducted the surveys. Based on the SCDNR records five additional wells were drilled during this period. The latest was drilled in 2002 in Jasper County to a depth of 700 ft. The deepest well in SC was drilled in 1984 in Colleton County to a depth of 12,700 ft. More work is needed by first obtaining and then re-examining the log data acquired in the wells to indicate the possible presence of hydrocarbon reservoirs that may have been overlooked. Further research is needed to determine the nature and extent of the oil and gas shows reported in the exploration wells drilled in South Carolina.

Bibliography

Reid JC (2009) North Carolina geological survey, Information circular 36, Natural Gas and Oil in North Carolina

South Carolina (2008) Reasonably foreseeable development scenario for fluid minerals, U.S. Department of the Interior

Chapter 12
Regional Geology, Oil and Gas Shows of Eastern Virginia, North Carolina, South Carolina and Middle Florida

Regional geology highlights indicate potential in Cretaceous and Paleozoic rocks. The Coastal Plain and Piedmont of the southeast United States are under-explored. Stratigraphic overlaps due to progression and regression of the sea probably produced strand line stratigraphic traps easily missed by earlier exploration methods of sparse drilling and low-resolution surface 2D seismic data profiling. There is strong evidence of some large untested structures near and in the Piedmont ahead of the Appalachian Mountains including northeast Georgia imaged by the COCORP seismic survey. The prospective formations are the Vicksburg, Wilcox, Ripley, Eutaw and Tuscaloosa carbonates and sandstones. Also, the Floyd and Conasauga shales and conglomerate sequences from the eroding Appalachian Mountains are prospective. In Georgia, oil and gas shows were reported in Miocene, Wilcox and Eutaw sands. Several oil seeps are known to exist and others have been reported. Over 80,000 sq. miles of sparsely drilled real estate exist in Georgia and Florida alone. Oil and gas shows were encountered in a central North Carolina well in recent years located in Lee County. Gas shows were reported in a well drilled in Mathews County, Virginia in the 1920's near the seacoast. Coal exists in northwest Georgia and was mined during the Civil War and again as recently as the 1970's – 1980's. There are also coal bed natural gas (CBNG) possibilities. Fig. 4.1.

Referring back to Fig. 4.1 on page 49 shows the on-shore part of the East Coastal Basin complex with long and narrow rift basins with the larger features such as the Deep River Basin in North Carolina and the Taylorsville Basin in Virginia and Maryland. The large and extensive South Georgia Basin and the North Florida Basin features may be actually one large basin.

Small oil and gas seeps have been observed in the South Georgia Basin for many years and appear to follow a northeast to southwest trend along the Fall Line in Georgia. The entire Fall Line region that extends for many hundreds of miles from the states of Alabama and on up to New Jersey should be closely re-examined for the evidence of hydrocarbon seeps and methane gas escape measurements that may point the way to hydrocarbons trapped in the subsurface. Fig. 2.1.

R. J. Brewer, *Hydrocarbon Potential in Southeastern United States*,
SpringerBriefs in Earth Sciences, https://doi.org/10.1007/978-3-030-00218-3_12

The basins shown in Fig. 4.1 were first delineated at some point with the help of surface geology mapping sources and limited well log data and are woefully under-explored pertaining to hydrocarbon accumulation potential. More conclusive information about each basin's extent, depth and structural make up will be gained when more regional seismic data is obtained and mapped and more exploratory wells are drilled.

Bibliography

Baum RB (1953) Oil and gas exploration in Alabama, Georgia and Florida, AAPG, pp 340–359
Mississippi Geological Society (AAPG) (1941) Possible future oil provinces of the United States
Reid JC (2009, March) Natural gas and oil in North Carolina. North Carolina Geological Survey, Information circular 36

Chapter 13
The Historic Jay and Sunniland Oil Fields of Florida

As of this writing there are only two oil-producing areas in Florida. One is in far southern Florida near the Everglades, and comprised of 14 separate fields, and the other is in the far western part of the state in the panhandle comprised of seven separate fields. The south Florida fields are located in Lee, Hendry, Collier, and Dade Counties. Florida's first oil field is the Sunniland field located in Collier County and was discovered in 1943. It has produced over 18 million barrels of oil. Later, 13 more field discoveries were made and lie along a northwest-southeast trend through Lee, Hendry, Collier, and Dade Counties. Although the fields are relatively small, production is significant. Together, the three Felda fields (West Felda, Mid-Felda, and Sunoco Felda) in Hendry County have produced over 54 million barrels of oil (Fig. 1.1).

The south Florida fields produce oil from small "patch reefs" within the Lower Cretaceous Sunniland Formation from between 11,500 and 12,000 feet. The Sunniland pay zones vary from about 5 to 30 feet in thickness.

The depositional environment during the Lower Cretaceous in south Florida was one of a shallow sea with a very slowly subsiding sea bottom. The time interval was characterized by numerous transgressions and regressions of the sea over the land which created a carbonate-evaporite sequence of geologic formations. It is believed that the Sunniland "reefs" are not true patch reefs but were localized mounds of marine animals and debris on the sea floor that lithified and became an oil and gas reservoir sequence. The primary mound-building invertebrates found in the Sunniland limestone were rudistids - oyster-like mollusks that existed only during the Cretaceous. They lived in great profusion and were widely distributed in clear, shallow Cretaceous seas. Other marine life found fossilized within drill cores, in the Sunniland patch reefs, or mounds, included calcareous algae, seaweed, foraminifera, and gastropods - such as snails.

Foraminifera, usually quite small, are single-celled animals with external skeletons or tests. Because of their seemingly incalculable large numbers in the seas, their tests and remains can represent significant amounts of organic debris eventually

R. J. Brewer, *Hydrocarbon Potential in Southeastern United States*,
SpringerBriefs in Earth Sciences, https://doi.org/10.1007/978-3-030-00218-3_13

settling to the ocean bottom. Pellets and other organic debris also accumulated in these mounds. The remains of the rudistids and other marine life and debris were deposited on the sea floor, forming porous limestones. Porosity within the limestones was enhanced over succeeding eons by the gradual transformation of limestone to dolostone, which resulted in good reservoir rocks to contain the oil.

The porous limestones and dolostones grade laterally into non-porous, chalky lime mudstones. These dense limestones form a barrier to oil migration, thus trapping the oil in the more porous rocks. Research indicates that the dense mudstones are probably the source rocks for the Sunniland oil. The Sunniland Formation, therefore, appears to include its own oil source rocks and some of its own seals. Additional seals are provided by the evaporites of the overlying Lake Trafford Formation.

On the other side of the state way to the northwest of the Everglades production in the western panhandle of Florida began with the discovery of Jay field in June 1970. Most people may not know that Jay is the largest oil field discovered in North America on land since the discovery on the Alaskan North Slope of the giant Prudhoe Bay field in 1967.

The Prudhoe Bay field was not the first oil and gas field discovered in Alaska. The first major field discovered in Alaska was on the Kenai Peninsula at Swanson River in 1957 in far southern Alaska hundreds of miles from the North Slope. Prior to the Swanson River discovery, oil seeps had long been observed in Alaska. A tremendous amount of oil still remains to be developed in the North Slope area between the large fields of Prudhoe Bay and Kuparuk (discovered in 1969) on the west and the long-disputed Point Thomson (discovered in 1975) on the east that includes the Arctic National Wildlife Area (ANWR) in between and other areas both onshore and offshore Alaska including the Chukchi Sea.

A very large natural gas cap accumulation still remains to be exploited from Prudhoe Bay and awaits adequate gas pipe line facilities to help get it to domestic and international markets that may involve the Keystone Pipeline project far to the south in the Lower 48 United States. Plans are in active development and being led by the state of Alaska Department of Natural Resources to bring Prudhoe Bay natural gas to markets involving LNG (Liquified Natural Gas) projects.

Since the discovery of the Jay Field, an additional six oil fields have been discovered in the western panhandle of Florida. The pay zones of the fields are from about 14,500 to 16,800 feet and vary in thickness from about 5 to 259 feet.

North Florida has dominated Florida oil production since the discovery of Jay field. North Florida oil fields accounted for 83 percent of the state's cumulative production through January 1988. The Jay field alone is responsible for 71 percent of the state's cumulative production and has recoverable reserves conservatively estimated in 1974 to be approximately 337 million barrels of oil. More reserves may be recoverable when more secondary and tertiary recovery methods are employed in oil production efforts including water and CO_2 gas flooding and injection. A known fact involving petroleum engineering is that it is not possible with current technology to recover all of the hydrocarbons held in any oil and gas reservoir rock.

Hydraulic fracturing of a hydrocarbon bearing rock formation using mostly water and sand under pressure is an established method used globally of maximizing the amount of oil and gas that a formation can ultimately produce. Hydraulic fracturing is not a new technology and was reportedly first used commercially in 1949 by Halliburton on a project on Oklahoma. Fracturing a reservoir rock hydraulically this way can allow an operator to reportedly get as much as 30% more oil and gas out of a reservoir.

The Jay field is located within the "Jay trend" of Escambia and Santa Rosa Counties in Florida, and Escambia County, Alabama. The Jay trend produces oil from Jurassic age Smackover formation carbonates and Norphlet Sandstones. In Florida, the Jay fields include Jay, Mt. Carmel, Coldwater Creek, and Blackjack Creek. The Jay trend fields in Florida and Alabama are associated with a normal fault complex which rims the Gulf Coast and is believed to extend to the south-southwest and far under the Gulf of Mexico waters.

The other Florida panhandle oil fields are Bluff Springs, McLellan, Sweetwater Creek, and McDavid. The Bluff Springs field probably formed as the result of a structure created by movement of the underlying Louann Salt. McLellan and Sweetwater Creek are probably associated with small salt structures or with the stratigraphic pinch out of the Smackover Formation.

Production for all of the Florida panhandle oil fields, except Mt. Carmel, is from Jurassic age Smackover dolostones and limestones. The Mt. Carmel field produces from both the Smackover and the underlying Jurassic age Norphlet Sandstone. Although a mixture of carbonates and clastics can be found within the Smackover in the western panhandle producing area, it is almost purely a sequence of dolostones and limestones. The underlying Norphlet Sandstone is primarily an arkosic or feldspar and quartz rich sandstone. The Norphlet is underlain by the Louann Salt. The Smackover Formation is overlain by the Buckner Member of the Haynesville Formation. The Buckner is composed of anhydrite, and other evaporites, and forms the seal to some of the Smackover producing zones.

The Sunniland oil field trend is located in Lee, Hendry, Collier and Dade Counties in far southern Florida near the Everglades. The area is quite unique and is probably one of the most isolated oil fields in the United States because it is many miles from other established onshore producing areas not including existing oil and gas discoveries made offshore in the eastern Gulf of Mexico and most interestingly have yet to be exploited to alleviate the country's oil and gas dependence. The Sunniland trend is isolated; however, from the standpoints of geology, sediment deposition and stratigraphy, it is probably part of the Gulf of Mexico geosyncline and basin complex. The trend is somewhat similar the substantial and widely known oil and gas production found west ward from the Jay field of the Florida panhandle on to Alabama, then Louisiana and continuing on to Texas and is not considered to be part of the East Coastal basin.

Additional oil and gas accumulations are believed to exist in the eastern Gulf of Mexico near Florida; however, are currently off-limits due to the existence of a drilling moratorium imposed by the U.S. government in the early 1990's. The Sunniland oil field was discovered by the Collier family in 1943 after in part being encouraged by Federal government incentives to find new oil supplies during World War II. The Sunniland oil field trend produces primarily from a tidal shoal sequence of Cretaceous age detrital carbonates at a depth of approximately 11,200 feet. The crude oil has been described as "immature" 21–28° API with a gas to oil ratio (GOR) of 85:1. Eight (8) separate fields make up the Sunniland producing trend and include: West Felda (largest), Mid-Felda, Sunoco-Felda, Bear Island, Corkscrew, Lehigh Park, Raccoon Point, and Sunniland. As of 1993, the Sunniland field complex had produced about 103 MMBO. 370 MMBO and 57.5 BCFG (6 MMBOE) are estimated to remain to be discovered and produced in the state of Florida. Six (6) oil & gas exploration plays identified by a USGS study: Upper Sunniland, Tidal Shoal, Lower Sunniland Fractured "Dark Carbonate", Dollar Bay Shoal – Reef Dolomite, Lower Cretaceous Carbonate Composite, Extended Upper Sunniland Tidal Shoal and Wood River Dolomite Deep Gas.

Bibliography

Lane E (ed) (1994) Florida's geological history and geological resources. Special Publication No. 35. Florida Geological Survey, Tallahassee

Pollastro RM (1995) USGS National oil and gas play–based assessment of the South Florida Basin, Florida Peninsula project, 2001

Sweet JM (2008) Discovery at Prudhoe Bay. Hancock House Publishers, Surrey, 304 pages

Chapter 14
Proposed Regional Reconnaissance 2D Seismic Survey Program

A reconnaissance 2D reflection surface seismic grid of very long lines of at least 50 miles each is needed to high grade existing drilling prospects and locate new drilling prospects. 3D surface seismic is not as useful for reconnaissance and regional work. 3D seismic surveys usually cover much smaller land areas and are considerably more expensive than 2D data. 3D surface seismic coverage will become applicable once new and commercially viable field discoveries are made and reservoir economics are considered and developments wells are being planned. The 2D surface seismic that appears to exist in the southeast United States is low resolution and fold and very limited in total individual line length. The surveys were typically designed and acquired for drilling prospect delineation by medium and large size operators and oil and gas exploration companies that have explored and drilled in the region. An exception is the COCORP research program mentioned in Chap. 8. Most of the oil company data is proprietary and not part of the public domain; although some of it, the majority of which being 2D, has probably been released by the owners to the market and is available through seismic data brokers for a license fee per linear or square mile (Figs. 14.2, and 14.3).

Figure 14.2 is a view of COCORP line 11 dominant dip vectors. Line 11 is a north to south 2D seismic line profile located in southern Georgia. See Fig. 14.1. It is an unmigrated seismic profile image meaning that the reflection points displayed on it were not moved or "migrated" back to their calculated point of origin in the subsurface. In practice, unmigrated seismic data is typically easier to interpret and map than migrated seismic data. The difference relates to seismic travel time and can result in seismic time to depth conversion errors if unmigrated data is mixed with migrated data within the same mapping project. Migrated seismic data displays are especially useful in viewing folded and faulted features identified in the data that tend to be not as clearly discernable on unmigrated seismic data.

COCORP line 11 is an example of the seismic data quality that was obtainable in part of the southeastern United States by the best equipment available in the early

© The Author(s), under exclusive licence to Springer Nature Switzerland AG 2019   71
R. J. Brewer, *Hydrocarbon Potential in Southeastern United States,*
SpringerBriefs in Earth Sciences, https://doi.org/10.1007/978-3-030-00218-3_14

Fig. 14.1 Traverse of COCORP seismic lines in southern Georgia and northern Florida

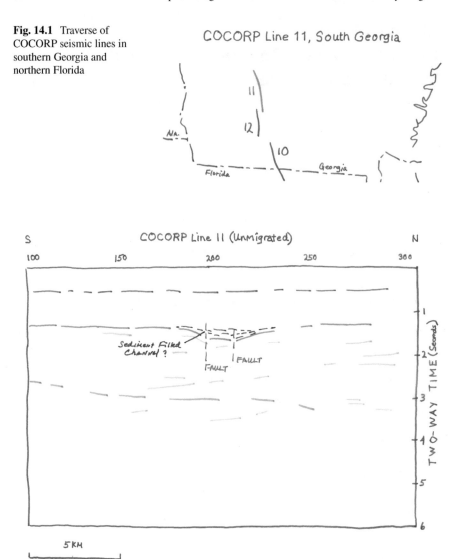

Fig. 14.2 COCORP dominant dip vectors profile line 11 in southern Georgia

1980's. It is expected that new seismic data acquired in the area will be of significantly improved quality allowing for more conclusive subsurface mapping.

Figure 14.3 is a map using an interstate road map as a base and shows an extensive proposed traverse of numerous 2D seismic lines that if ever acquired will yield a quite unique comprehensive and valuable base of seismic coverage. The line traverses are planned along roads and some cases railroads as paths of least resistance. The surveys will allow for the first time accurate mapping of the subsurface in the

Fig. 14.3 Proposed regional reconnaissance 2D seismic survey program Approximately 7320 line miles. Estimated data acquisition, permitting and data processing cost: $25,000/mile or $183 million

southeastern United States and tie many key well locations together. Features and anomalies will no doubt be identified and mapped and some will warrant further investigation and testing by drilling.

The ideal seismic grid will also extend farther to the north and west up and over the Appalachian Mountains and extend well into the Appalachian and Black Warrior Basins to tie key well locations. Interpretations and mapping projects will produce for the first time significantly clearer understandings of the structural geology and stratigraphy between the East Coastal Basin and the Appalachian Basin that up until now have only been barely possible with very sparse well log data and limited surface geology mapping information and have yielded very limited information about the subsurface.

Ambitious seismic data acquisition projects in difficult terrain more challenging than the southeastern United States are now possible with the latest seismic data acquisition and processing technology. Great achievements have been made by the geophysical data industry in equipment portability, reliability and environmental sustainability. Similar projects have been undertaken and successfully completed in other parts of the world including South America and Asia.

Seismic data brokers also sell available 3D seismic data that is typical priced on a per square mile basis. A significant portion of formerly proprietary 2D and 3D seismic data may also be available on an outright purchase basis where the buyer becomes the new owner of the data set as opposed to the usually less expensive license fee alternative. The data brokerage and sale of formerly proprietary data owned by oil and gas exploration companies is a long established and successful business model that was conceived and started by a few innovative American businessmen some of whom themselves were former oil company employees. The seismic data brokerage companies later become involved in acquiring their own seismic data coverage on a speculative basis and underwritten by their oil company customers in turn for an exclusive use period after which the data brokerage/speculative data firm could repeatedly re-license the data later to other oil company customers. The data brokerage and speculative seismic data business started in earnest in the early 1970's and possibly even earlier.

There is sparse shallow well control and few tests over 5000 ft. One of the deepest tests in Georgia is to a TD of 11,470 feet drilled by Chevron in 1981. Very limited well log and formation testing and evaluation data exists and is of older technology and low resolution. Some potentially productive reservoirs may have been overlooked. There has been very little recent drilling. Structural and stratigraphic anomalies remain untested in northern Georgia, northern Florida, South Carolina, North Carolina and eastern Virginia. A dynamite shot hole energy source is preferable with vibrator truck in-fill in high culture areas. Heli-portable operations are an option in rugged terrain in the Piedmont near the Appalachian mountain fronts. Nodal cable- less telemetry data acquisition equipment is state of the art and is an option that will be considered. Long 2D seismic lines are needed to tie the distant producing field areas together and extend to the southeast to prospective areas. The possibilities include: Alabama to Florida, Alabama to Georgia, Tennessee to Georgia, Tennessee to South Carolina, Tennessee to North Carolina, Kentucky to North Carolina, Ohio and West Virginia to Pennsylvania to Maryland and Virginia. A notable fact is that many wells in Georgia were drilled near railroads probably because of easier land permit and surface access logistics. The use of state highway roads and railroads is possible to maximize traverse access and minimize permit and land entry costs. Technology advancements to help minimize the substantial costs of new seismic data acquisition will be very helpful. The use of Landsat and other satellite based remote sensing and imaging technologies will help save time and reduce costs in efforts to high grade subsurface formation target areas.

The sonic log from the Chevron A.P. Snipes, et al., No. 1 location in Jeff Davis County, Georgia is the deepest well drilled to date in the state. The sonic log should make a very useful synthetic seismogram from 7500 ft. to the TD of 11,470 ft. and be an effective formation correlation tool to the surface seismic data that ties the Chevron well and other seismic lines nearby. The sonic log from the nearby, just to the north and much shallower (4063 ft.), Chevron J.L. Sinclair No. 1 also in Jeff Davis County, Georgia will help with making a synthetic seismogram from 4063 ft. to the near surface.

Most of the exploration wells drilled in Georgia reached only to the first section of igneous crystalline rock in the 4000 ft. to 6000 ft. total depth range. A very few wells like the Chevron Snipes location went deep enough to indicate sedimentary rock sections sandwiched in between two or more massive sets of crystalline igneous rock. The in between softer sedimentary or meta-sedimentary rock sections clearly appear on the sonic logs and continue to be prospective for possible oil and gas accumulations under the coastal plain.

Chapter 15
Offshore Southeastern United States and OCS Oil and Gas Potential

In the late 1970's the United States Atlantic coast was divided up into four (4) areas by the U.S. Department of the Interior (DOI) and include North Atlantic (1), Mid Atlantic (2), South Atlantic (3) and Gulf of Mexico (4).

Following extensive marine seismic surveys and then Outer Continental Shelf (OCS) land sales by the DOI, a total of 51 oil and gas exploration wells were drilled - most in water thousands of feet deep by major U.S. oil companies including Amoco and Shell off the United States Atlantic coast starting in 1978. No commercial quantities of oil and gas were found, although commercial quantities were discovered in 1971 much farther north on the Scotian Shelf (Sable Island/Cohasset) and in the Jeanne d'Arc Basin off the Grand Banks, Newfoundland (Hibernia) of Canada. Potentially commercial amounts of natural gas and some oil were also discovered in 1979 in Jurassic age carbonate and sandstone rocks under much shallower water about 400 feet deep nearer to the coast about 100 miles east of Atlantic City in key shelf wells drilled by Texaco, Tenneco and Exxon in the Baltimore Canyon area of offshore New Jersey. One well reportedly tested 18.9 million cubic feet of gas per day at a depth range of 13,000 to 15,000 ft.

Oil shows were found in wells drilled in the state waters of the Florida Keys during the time period from 1946 to 1962. The discoveries near the coast are now off-limits to further confirmation drilling as a consequence of a drilling moratorium imposed by the U.S. government in 1990 and later twice extended. In the southeast Georgia embayment of offshore Georgia, seven dry wells were drilled during the late 1970's to early 1980's time frame. Much farther to the south in the North Cuba Basin 3 producing fields have been discovered in the waters of offshore Cuba.

In spite of the discouraging industry track record; recent (2012) Bureau of Ocean Energy Management (BOEM) formerly known as the Minerals Management Service (MMS) petroleum resources estimates for the U.S. Atlantic OCS region are

3.3 billion barrels (BBO) of recoverable oil and 31.3 trillion cubic feet (TCF) of natural gas. The estimates remain quite substantial if albeit optimistic and should warrant further exploration drilling efforts when timing is better in the future and drilling moratoriums are not as widespread in U.S. waters.

Bibliography

Adinolfi F AAPG search and discovery, hydrocarbon potential of the Baltimore canyon trough, MMS, Herndon, VA

Erlich B (2014) OCS: potential doesn't always ensure success, AAPG Explorer, May 2014

Chapter 16
Prospective Areas for New Oil and Gas Exploration and Current and Future Developments

The Marcellus shale resource play is being actively drilled and developed in western Virginia and western Maryland. The Conasauga and Floyd shale sequences/equivalents exist in Georgia. The Floyd/Neal shale sequence is productive in Alabama. East Tennessee and northern Alabama produce oil and gas and need more exploration. The Chattanooga shale in Tennessee is a prolific oil source rock. Unconventional shale gas potential from the Chattanooga shale in Tennessee is a strong possibility.

The sedimentary basins on the eastern slope of the Appalachian Mountains in Virginia, North Carolina, South Carolina, Georgia and Florida are under-explored. The entire Fall Line region that extends for many hundreds of miles from the states of Alabama and on up to New Jersey should be closely re-examined for the evidence of hydrocarbon seeps and methane gas escape measurements that may point the way to hydrocarbons trapped in the subsurface.

The data interpretations using the COCORP 2D seismic survey (1980's) provide indications of sediments existing under over-thrusted igneous and metamorphic rocks. Some oil and gas accumulations in Georgia and North Carolina yet to be discovered may be in parts of the states that are near the borders with Tennessee and Alabama. Hydrocarbons found in shale formations may be exploited by proven established as well as new and improved horizontal drilling and hydraulic fracturing technologies that are more sustainable and will have less potential impact on the environment.

South Carolina, North Carolina, Virginia and Georgia in addition to being vastly under-explored on land also have encouraging offshore potential. The Florida onshore and offshore petroleum potential in the eastern Gulf of Mexico (GOM) is well documented and will benefit from more reasonable and proactive sustainability planning and environmental safeguards.

Cocorp line 11 traverses for approximately 14 miles extending north to south in southern Georgia. See Figs. 14.1 and 14.2. Line 11 is very unique within the COCORP data collection because it exhibits an interesting structural and stratigraphic anomaly that can be interpreted as a large buried river channel filled with sand and sediment that may serve as a large hydrocarbon trap. Available well logs near the anomaly are being examined in detail to identify rock formations and sedimentation mechanics and age. The anomaly seen on line 11 may in fact be several separate structural and stratigraphic features. Several more seismic lines traversing north south and east west are needed to delineate, establish four way structural closure and see if similar neighboring sediment channel anomalies are visible.

The anomaly is near the areas of two large regional structural features that have influenced sediment deposition in southern Georgia known as the Gulf Trough and the Chattahoochee Anticline. See Fig. 10.1.

Only drilling will positively confirm the channel feature as hydrocarbon bearing. The channel feature is estimated to be only a few thousand feet below the surface and well within the reach of modern drilling technologies.

Conclusions

More wells will need to be drilled in the area and logged using more advanced formation evaluation technologies including image logs, magnetic resonance imaging logs and vertical seismic profiles. Additional drilling prospects will result after a comprehensive and extensive grid of 2D higher fold reflection seismic data is acquired, processed and interpreted. The seismic data will be tied to key well logs and then finally mapped that will tie key stratigraphic and structural features and anomalies together for the first time on a larger and more meaningful scale. After oil and gas discoveries are made, higher resolution 3D reflection seismic data acquisition techniques can be employed to delineate the extent of the new fields and be used to locate development well locations and reduce the risk of drilling dry holes. Other established technologies that have been used in the area before and may prove useful yet again are modern and higher resolution gravity and magnetic data survey technologies that will remain attractive because of their lower cost compared to reflection seismic surveys. Remote sensing technologies including Landsat and other satellite based imaging technologies are expected to play an important and efficient role. Innovative and newer geochemical and surface methane measurement survey technologies are still being evaluated; however, may become increasingly useful and may help find hydrocarbon accumulations more quickly and at lower cost.

A potentially productive hydrocarbon fairway may exist between a trend line traced through all of the known oil seeps in Georgia and the Fall Line. The author will not be surprised if the first commercial discovery of oil and/or gas made in Georgia is within or near this fairway.

Additional oil seeps in Georgia may be discovered if the Fall Line area is studied more closely. The Fall Line and Coastal Plain are very large topographic features in the southeastern United States and need to be studied much more closely for clues to the presence of hydrocarbons.

© The Author(s), under exclusive licence to Springer Nature Switzerland AG 2019
R. J. Brewer, *Hydrocarbon Potential in Southeastern United States*,
SpringerBriefs in Earth Sciences, https://doi.org/10.1007/978-3-030-00218-3

The main purpose of the writing effort is to re-introduce and share information about a large and prospective region not widely known to most people including seasoned oil and gas industry professionals. Success will in part result in the stimulation of further discussion and most importantly specific action involving land leasing and well drilling. It is recommended, hoped and expected that oil and gas exploration work and investment continue in the southeastern United States. Continued effort will be part of a long-sought after comprehensive energy security strategy as our country slowly but steadily moves toward the ultimate goal of total crude oil and natural gas independence.

Suggested Additional Reading

Georgia

Cook FA, Brown LD, Sidney K, Oliver JE, Petersen TA (1981) COCORP seismic profiling of the Appalachian orogen beneath the coastal plain of Georgia, Geol Soc Am Bull, Part 1, v 92, p 738–748

Cooke WC, Munyan AC (1938) Stratigraphy of coastal plain of Georgia. AAPG Bull 22(7):789–793

Hurst VJ (1960) Oil tests in Georgia, Information Circular 19, Georgia State Division of Conservation, Department of Mines, Mining and Geology, p 14

Mondary JF, Pratt TL, Brown LD, A 3D look at the midcrustal surrency bright spot beneath Southeastern Georgia, CH5.7, p 279–282

Munyan AC (1938) Recent petroleum activities in coastal plain of South Georgia. AAPG Bull 22(7):794–798

Nelson KD, Arnow JA, McBride JH, Willemin JH, Huang J, Zheng L, Oliver JE, Brown LD, Kaufman S (1985) New COCORP profiling in the southeastern United States. Part 1: Late Paleozoic suture and Mesozoic rift basin. Geology 13:714–718

Georgia State Division of Conservation, Department of Mines, Mining and Geology (1965) Oil Tests in Georgia, Information Circular 19, 3rd edn, p 16

Patterson SH, Herrick SM (1971) Chattahoochie Anticline, Apalachicola Embayment, Gulf Trough and Related structural features, Southwestern Georgia, Fact or Fiction, The Geological Survey of Georgia Department of Mines, Mining and Geology, Information Circular 41, 16 pages

Prettyman TM, Cave HS (1923) Petroleum and natural gas possibilities in Georgia, Geological Survey of Georgia, Bulletin No. 40. Byrd Printing Company, Atlanta, p 164

Simmons MSC, Elizabeth F, Ziegler DG, Aiken CLV Integrated geophysical study of the Hawkinsville Triassic Basin, Georgia: problems in geophysical data interpretation in the South Georgia Rift, CS 1.4, p 12–15

U.S. Department of the Interior, BLM (2008a) Georgia, Reasonably foreseeable development scenario for fluid minerals, p 13

University of Georgia – Department of Geology (2005) The Geology of Georgia, p7

© The Author(s), under exclusive licence to Springer Nature Switzerland AG 2019   83
R. J. Brewer, *Hydrocarbon Potential in Southeastern United States*,
SpringerBriefs in Earth Sciences, https://doi.org/10.1007/978-3-030-00218-3

Mississippi

Mississippi Geological Society (1941) Possible Future Oil Provinces of the Southeastern United States. AAPG Bull 25(8):1575–1586

Maryland

Maryland Geological Survey (2011) Is there a natural gas boom in maryland's future?, Information Concerning the Marcellus Shale and the Search for Natural Gas in Western Maryland, p 2

Vermont

Varricchio L (2014) A history of oil and gas in vermont, The eagle.com, p 5
Vermont Geological Survey (2009) Vermont oil and gas exploration wells and water wells with gas shows, p 1
Vermont Geological Survey (2012) Oil and gas wells drilled in vermont, p 1

Virginia

Costain JK, Coruh C Tectonic setting of triassic half-grabens in the appalachians: seismic data acquisition, Processing and results, Chapter 11, pp 155–174
Virginia Department of Mines (2011) Minerals and energy, Virginia division of mineral resources, oil and gas, p 4

North Carolina

Coffey JC (1977) Exploratory oil wells of North Carolina 1925–1976, Information circular no. 22, North Carolina department of natural and economic resources, division of earth resources, geology and mineral resources section, p 52
Reid JC (2009) North Carolina geological survey, natural gas and oil in North Carolina, information circular 36, p 8
Reid JC, Milici RC (2008) Hydrocarbon source rocks in the deep river and Dan river Triassic basins, North Carolina, USGS and North Carolina geological survey, open-file report 2008–1108, p 27
Reid JC, Taylor KB, (2013) Mesozoic rift basins – onshore North Carolina and south-central Virginia, U.S.A.: Deep River and Dan River – Danville total petroleum systems (TPS) and assessment units (AU) for continuous gas accumulation. Open-file report 2013–01, North Carolina Geological Survey Section, p 8

South Carolina

Griffin WT (1989) Historical summary of onshore oil and gas exploration activities in South Carolina, South Carolina water resources commission, open-file report 30, p 7

U.S. Department of the Interior, BLM (2008b) South Carolina, reasonably foreseeable development scenario for fluid minerals, p 15

Tennessee

Berwind MB (1987) The history and development of the oil and gas industry in Tennessee. J Tenn Acad Sci 62(3):60–62

Alabama

Braunstein J (1953) Fracture-controlled production in Gilbertown field, Alabama. AAPG Bull 37(2):245–249

Florida

Herbert TA Lampl LL (2001) Oil and gas energy issues in Florida's future, Policy Report # 33, James Madison Institute, p 33

Lane (ed) (1994) Florida's geological history and geological resources, Special Publication No. 35, Florida Geological Survey, p 74

Lloyd JM (1989) 1986 and 1987 Florida petroleum production and exploration, Information Circular No. 106, Florida Geological Survey, p 39

Lloyd JM (1992) 1990 and 1991 Florida petroleum production and exploration, Information Circular No. 108, Florida Geological Survey, p 31

Lloyd JM (1994) 1992 and 1993 Florida petroleum production and exploration, Information Circular No. 110, Florida Geological Survey, p 30

Lloyd JM (1997) 1994 and 1995 Florida petroleum production and exploration, Information Circular No. 111, Florida Geological Survey, p 63

Mineral Information Institute (2005) Florida's industrial minerals, p 3

Temple B (2012) Sunniland shale – an emerging South Florida basin liquids play. Oil Gas J 110:5

Toulmin LD (1955) Cenozoic geology of Southeastern Alabama, Florida and Georgia. AAPG Bull 39(2):207–235

Oil and Gas Seeps

Aminzadeh F, Berge TB, Connolly DL (eds) (2013) Hydrocarbon Seepage: Source to Surface, Geophysical Developments No. 16. SEG and AAPG, Tulsa, p 244

Link WK (1952) Significance of Oil and Gas Seeps in World Oil Exploration. AAPG Bull 36(8):1505–1540

General

Appalachian and Illinois Basin Directors of the Interstate Oil and Gas Compact Commission (IOGCC) (2005) Mature region, youthful potential, oil and natural gas resources in the Appalachian and Illinois basins. Interstate Oil and Gas Compact Commission, Oklahoma City, p 28

Brewer RJ (2003) VSP is a Check Shot Step Up, AAPG Explorer, 2000, The Check Shot Velocity Survey, AAPG Explorer, 2000, The Look ahead VSP, AAPG Search and Discovery, 2002, Comprehensive Use of VSP Technology at Elk Hills Field, Kern County, CA., SEG, 2003

Dobrin MB, Savit CH (1988) Introduction to geophysical prospecting, 4th edn. McGraw-Hill Book Company, New York, p 867

Hubbard S, Sharpless C, Cahit C, John K (1991) Paleozoic and Grenvillian structures in the Southern appalachians: extended interpretation of seismic reflection data, American Geophysical Union. Tectonics 10(1):141–170

LeRoy LW, LeRoy DO (1977) Subsurface geology petroleum mining construction. Colorado School of Mines, Golden, p 941

Levorsen AI (2001) Geology of petroleum, A commemorative second edition, AAPG, p 724

Pohn HA (1998) Lateral ramps in the folded Appalachians and in overthrust belts worldwide – a fundamental element of thrust-belt architecture, USGS bulletin 2163. United States Government Printing Office, Washington, p 36

Rand McNally, Road Atlas, United States, Canada and Mexico 2015, p 140

SEG, USGS, DMA and NOAA (1982) Gravity anomaly map of the United States, east half. The Society, Tulsa

Sherriff RE (2002) Encyclopedic dictionary of exploration geophysics, 4th edn. SEG, Tulsa, p 429

Index

Printed in the United States
By Bookmasters